# Tidal Deformations on a Rotating, Spherically Asymmetric, Visco-elastic and Laterally Heterogeneous Earth

# European University Studies

Europäische Hochschulschriften

Publications Universitaires Européennes

**Series XVII**

**Earth Sciences**

Reihe XVII Série XVII

Geowissenschaften

Sciences de la terre

**Vol./Bd. 5**

PETER LANG

Frankfurt am Main · Bern · New York · Paris

Rongjiang Wang

# Tidal Deformations on a Rotating, Spherically Asymmetric, Visco-elastic and Laterally Heterogeneous Earth

PETER LANG
Frankfurt am Main · Bern · New York · Paris

Die Deutsche Bibliothek - CIP-Einheitsaufnahme

Wang, Rongjiang:

Tidal deformations on a rotating, spherically asymmetric, visco-elastic and laterally heterogeneous earth / Rongjiang Wang. - Frankfurt am Main ; Bern ; New York ; Paris : Lang, 1991

(European university studies : Ser.17, Earth sciences ; Vol. 5)
Zugl.: Kiel, Univ., Diss., 1991
ISBN 3-631-43991-1

NE: Europäische Hochschulschriften / 17

D 8
ISSN 0721-5045
ISBN 3-631-43991-1

# Contents

## Abstract

Perturbation method is applied to determine tidal deformations on a realistic earth. The earth's rotation, lateral heterogeneities, hexagonal anisotropy and visco-elastic dispersion of elastic moduli in the earth's mantle, spherically asymmetric configurations of the earth's surface as well as the interior boundaries, and deviations from the initial static equilibrium are considered as perturbations of first order. For perturbations of long spatial wavelengths, we make use of expansion series of spherical harmonics for both the tidal observables and the disturbed model parameters. A principal advantage of the perturbation method compared to the direct method is that the expansion coefficients of the solution can be evaluated independently from each other by a semi-analytical procedure. The validity of our approach has been checked against the analytical solutions for several special cases.

For the tidal motion on an elliptical, rotating and elastic earth we find some important discrepancies between the results of the present perturbation theory and those given in previous publications. The latitude dependences of the equivalent Love number k and the gravimetric factor $\delta$ are only half as large as those given by the previous results. By adopting the modified definition of k and $\delta$ suggested by Dehant (1987), in which the theoretical tides are related to the elliptical surface instead of the spherical one, there is, to first order, no latitude dependence of these two tidal parameters any more. However, the gravimetric factor shows an increase of 0.2%, 0.4% and 0.3% for the long period, diurnal and semi-diurnal tides, respectively, compared to the average values given by Dehant and Zschau (1989).

Like the earth's rotation and ellipticity, mantle heterogeneities also modify the tidal response of the earth. In addition to all effects of the rotation and ellipticity, the equivalent tidal parameters become complex and station dependent even in the purely elastic case because of the axial asymmetry. Perturbations in the Love numbers h and k are relatively sensitive to large-scale lateral heterogeneities in the shear modulus of the upper mantle. The effect on the gravimetric factor is systematically smaller. This is because the Love numbers h and k enter in the gravimetric factor with opposite sign and their perturbations cancel to a large extent. It is shown that mantle heterogeneities of all long wavelengths strongly effect the Love number l.

By using the theory of temperature induced relaxation and dispersion of the shear modulus in the mantle, it is suggested that the velocity-heterogeneities observed with seismic tomography should be considerably amplified at tidal periods if they are essentially induced by differences of the ambient temperature. Based on Zschau's rheological model (see Appendix D), the visco-elastic amplification factor is about 3-5 and 5-10 in the semi-diurnal band. This assumes the seismic reference period to be 200s and 30s, respectively. This order of magnitude of the visco-elastic amplification is a conservative estimate. In fact, it can be much stronger, if the reference period is taken to be the average period of the body waves which are used in seismic tomography. Consequently, the influences of mantle heterogeneities on earth tides increase significantly. With a rough estimation of the perturbations of density, the bulk modulus and the unrelaxed shear modulus, we find that for the semi-diurnal tide M2, the residuals in the vertical displacement and in gravity are intensified by a factor of 5-7 and reach up to the level of 1 mm and 200 ngal, respectively. In the horizontal displacement the residuals are increased by a factor of 3-4 and are at the level of 2 mm, and in the Eulerian potential they are increased by a factor of about 2 and reach up to 30 gal·cm, respectively. These effects are solely due to very large-scale lateral heterogeneities (up to degree and order 8), which corresponds to several thousand kilometers. If the heterogeneities of shorter wavelengths can be included and the seismic reference period for determining the visco-elastic amplification is taken to be much shorter than 200s, the total effects on the complete earth tides may be much larger.

## Zusammenfassung

Es wurde die Störungsmethode zur Bestimmung der Gezeitendeformationen einer realistischen Erde angewendet. Die Abweichungen von einem kugelsymmetrischen und elastischen Modell, z.B. durch die Erdrotation, durch laterale Inhomogenitäten, hexagonale Anisotropie und viskoelastische Dispersion im Erdmantel sowie die Oberflächen- bzw. innere Grenzflächenundulationen wurden als Störungen erster Ordnung behandelt. Im Bereich von langen räumlichen Wellenlängen wurden die Störungen der Gezeitenvariablen und die o.g. Parameterstörungen in Kugelfunktionen entwickelt. Die Aufgabe besteht dann darin, die Entwicklungskoeffizienten der Gezeitenstörungen des gestörten Systems zu bestimmen. Das Variationsprinzip und die Greensche Formel führen das Problem in eine Integralgleichung über und die Entwicklungskoeffizienten der Gezeitenstörungen können unabhängig voneinander durch die Volumen- bzw. Flächenintegrale berechnet werden. Die Gültigkeit der Theorie wurde in mehreren speziellen Fällen durch analytische Ergebnisse bestätigt.

Bei der Bestimmung der Effekte von Rotation und Elliptizität zeigten die Ergebnisse der vorliegenden Theorie signifikante Diskrepanzen zu den bisherigen Ergebnissen. Die Breitenabhängigkeit der äquivalenten Loveschen Zahl k und des äquivalenten Gravimeterfaktor $\delta$, so wie sie traditionell definiert wurden, ist nur halb so groß wie die von Wahr (1979) berechnete. Die äquivalenten Gezeitenparameter wurden von Dehant (1987) als die Verhältnisse zwischen den beobachteten Gezeitengrößen und den entsprechenden theoretischen Größen in dem selben Punkt an der elliptischen Erdoberfläche statt an einer Kugelfläche neu definiert. Mit dieser neuen Definition verschwindet die Breitenabhängigkeit der Loveschen Zahl k und des Gravimeterfaktors $\delta$ in erster Größenordnung von der Elliptizität. Für das gleiche Erdmodell zeigt der breitenunabhängige Gravimeterfaktor eine Zunahme von 0.2%, 0.4% bzw. 0.3% bei den langperiodischen, den ganztägigen bzw. den halbtägigen Gezeiten im Vergleich zu dem Mittelwert von Dehant und Zschau (1989).

Die Ergebnisse der seismischen Tomographie zeigen, daß der Erdmantel lateral inhomogen ist. Für die Raumwellen-Geschwindigkeiten liegen die lateralen Inhomogenitäten bei einigen Prozent. Zusätzlich zu Effekten aufgrund von Rotation und Elliptizität ergeben sich als Konsequenz der lateralen Inhomogenitäten die Stationsabhängigkeit und Phasenverschiebung der äquivalenten Gezeitenparameter. Die Störungen in den Loveschen Zahlen h und k sind relativ empfindlich bezüglich der großräumigen lateralen Inhomogenitäten des Schermoduls im oberen Mantel und zeigen ein ähnliches Verhalten. Die Störungen des Gravimeterfaktors sind systematisch kleiner, weil die Loveschen Zahlen h und k mit unterschiedlichen Vorzeichen in dem Gravimeterfaktor $\delta$ auftreten und sich ihre Effekte nahezu eliminieren. Die Störungen der Loveschen Zahl l zeigen keine deutliche Abnahme mit fallender Wellenlänge von den lateralen Inhomogenitäten im Mantel. Ihr prozentualer Effekt ist sehr groß.

Es wurde gezeigt, daß die lateralen Inhomogenitäten bei Gezeitenperioden wegen der viskoelastischen Relaxation und Dispersion des Schermoduls viel stärker sein können als bei seismischen Perioden, wenn sie vorwiegend durch Temperatur-Differenzen induziert werden. Nach dem rheologischen Modell von Zschau kann der sogenannte Verstärkungsfaktor für die halbtägigen Gezeiten in der Größenordnung von 3-5 bzw. 5-10 liegen, wenn die seismische Bezugsperiode von 200 s bzw. 30 s angenommen wird. Als Konsequenz wird der Einfluß lateraler Inhomogenitäten auf Erdgezeiten stärker als er im rein elastischen Fall erwartet wird. Für die halbtägigen Gezeiten M2 werden die Störungen in der vertikalen Verschiebung und in der Schwere um den Faktor 5-7 bis auf 1 mm bzw. 200 ngal verstärkt. In der horizontaler Verschiebung werden die Störungen um den Faktor 3-4 bis auf 2 mm und in dem Eulerschen Potential um den Faktor 2 bis auf 30 cm·gal verstärkt. Diese Effekte sind lediglich die Einflüsse der großräumigen lateralen Inhomogenitäten (bis auf den Entwicklungsgrad 8), deren Wellenlängen bei mehreren tausend Kilometern liegen. In den Modellrechnungen wurden die lateralen Inhomogenitäten der Dichte und der unrelaxierten elastischen Modulen grob durch die Anomalien seismischer Wellengeschwindigkeiten abgeschätzt. Die viskoelastische Verstärkung der Anomalien im Schermodul wurde für die Bezugsperiode von 200 s berücksichtigt. Wenn die kürzere Bezugsperiode der für die seismische Tomographie benutzten Körperwellen angenommen wird und die Inhomogenitäten von kürzeren Wellenlängen zusätzlich berücksichtigt werden, können die gesamten Effekte insbesondere in der horizontalen Verschiebung erheblich größer sein.

## Chapter 1. Introduction

As early as more than a century ago, Kelvin found that the equilibrium ocean tide is by about one of three parts lower than that expected by assuming the earth to be rigid. He explained this phenomenon by assuming the similar periodic displacement of the solid earth's surface and gave in 1883 a theoretical model describing the elastic deformations of an incompressible, homogeneous, spherical earth as a response to the lunisolar attraction. The observed equilibrium tide can be explained by his simple model if the shear modulus is chosen as that of steel. This is the earliest theoretical investigation on earth tides. At the beginning of this century, Love (1909) and Lamor (1909) modelled the tidal deformation on a radially stratified earth model.

With the improvement of traditional terrestrial techniques and the development of modern space techniques especially during the last decades, studies on earth's tides have become more important in the sense that, on one side, they are a useful geophysical signal for our understanding of the interior structure of the earth and, on the other side, they are a source of noise to be corrected for many other geophysical measurements. Since the fifties, a number of special subjects have been taking shape in earth tide research, for example, tidal measurements by using space techniques and terrestrial instruments, body tides, or traditionally called earth tides and ocean loading effects, core resonance and tidal nutation, mantle rheology and heterogeneity, and so on.

With respect to theoretical studies tidal motion has been modelled more realistically for an elastic, self-gravitating earth based on the theory of free oscillations (Alterman et al., 1959; Takeuchi et al., 1962; Longman, 1962, 1963; Backus, 1967; Gilbert & Backus, 1968; Wiggins, 1968; Farrell, 1972; Dahlen, 1972; and others).

The effects of the rotation and the elliptical shape of the earth, which were already mentioned by Love (1911) and estimated to be at the level of the ellipticity based on a homogeneous, incompressible model, were extented by including a fluid core and an elastic stratification throughout the earth (Jeffreys & Vicente, 1957a, b; Molodenskij, 1961; Sasao et al., 1980). In these studies, however, mantle deformation is still computed for a spherical, non-rotating shell. The results showed a significant effect for nearly diurnal tides due to the fluid core resonance. Unlike earlier theories, Wahr's (1979, 1981a, b, c, 1982) theory takes into account elliptical and rotational effects throughout the whole earth. Based on the previous works of Smith (1974, 1976, 1977), he computed the tidal deformations on an elliptical, rotating, elastic and oceanless earth by integrating a large set of equations of motion using the generalized spherical harmonic functions (Edmonds, 1960; Phinney & Burridge, 1973). The main differences of this model with respect to a non-rotating, spherical model are:

— *Latitude dependence of tidal parameters*: The tidal deformation can no longer be described by only the main term, which has the same degree and order as the tide-generating potential. There are small spherical harmonic terms of different degree. This effect of splitting of the degree can be represented by using equivalent tidal parameters which are thus latitude dependent;

— *Coupling of spheroidal and toroidal modes*: Spheroidal and toroidal deformations are no longer decoupled, but are coupled subject to specific coupling rules;

— *Order dependence of tidal parameters*: Although there is no splitting of the order because of axial symmetry, the equivalent tidal parameters are dependent on the order of the spherical harmonic term, i.e. they are different for the sectorial, tesseral and zonal tides;

— *Resonance in the diurnal band*: The equivalent tidal parameters in the diurnal band are strongly dependent on frequency. This effect is known as the nearly diurnal free wobble which is due to the resonance of the liquid core with aspherical core-mantle boundary;

It is, at present, difficult to experimentally verify these small effects because of other insufficiently modelled influences such as the effect of ocean tidal loading, calibration errors etc. In spite of this, Wahr's model of earth tides has been used recently as the a priori standard for modelling satellite tracking data (Marsh et al., 1988). Thus, it seems to be useful to solve the problem again by an independent method.

An important aim of earth tide studies is to contribute to our understanding of the mantle inelasticity, because tidal deformation covers the period range from a few hours to teens of years and exhibits two significant influences of the mantle inelasticity (Zschau, 1979, 1983; Zschau & Wang, 1986):

— *Phase shift*: The tidal response of the earth shows phase shifts to the tide-generating force;

— *Frequency dependence*: The tidal parameters are frequency dependent due to dispersion of the Love numbers.

At present, there exist many rheological models to describe the mantle inelasticity in the seismic band, for example, the so-called absorption band model with a flat or slightly inclined dependence of the quality factor, $Q \sim \omega^{\alpha}$, over an arbitrary finite frequency range (Liu et al., 1976; Kanamori & Anderson, 1977; Anderson & Minster, 1979; Minster, 1980; Müller, 1983). Recently, Zschau (Zschau & Wang, 1986, see also Appendix D) constructed a new rheological model based on a large set of data over a frequency range of more than seven decades including the seismic band as well as free oscillations and the Chandler wobble. This model supposes a generalized Maxwell rheology with a cut Gaussian distribution of the (logarithmic) stress relaxation times. In particular, it accounts for the influence of composition, structure, phase transitions, pressure and temperature on the inelastic behaviour in the earth's mantle by making use of an empirical relationship which is shown to be well fulfilled in various solids. Furthermore, it predicts the temperature induced dispersion and relaxation and allows us, as will be seen below, to obtain constraints on mantle heterogeneities from tidal tomography.

From seismic tomography it is known that the earth's mantle is laterally inhomogeneous to a few per cent in the seismic velocities. A global three-dimensional model has been recently constructed for the S velocity in the upper mantle and P velocity in the lower mantle (Dziewonski, 1984; Woodhouse & Dziewonski, 1984). Like the earth's ellipticity, mantle heterogeneities would modify the tidal response. In addition to all effects of ellipticity, the equivalent tidal parameters become

— *station dependent*, because there is not only a splitting of the degree but also a splitting of the order. For example, for tesseral terms in the tide-generating potential there may result sectorial and zonal terms in the tidal response, and the tidal parameters consequently depend upon latitude as well as longitude; and the parameters become

— *complex*, because a phase shift in the tidal response may be caused by the asymmetric distribution of the rheological parameters.

The exact magnitudes of these effects are not known. Molodenskij & Kramer (1980) estimated the effect in the gravimetric factor to a few per mill. In their model the seismic velocity in the suboceanic mantle is assumed uniformly to be by 5% smaller than that under the continent. But they did not take into account perturbations in the density, in the gravity field as well as in the initial static equilibrium, assuming these effects to be small. From the fact that the magnitude of the observed mantle heterogeneities on average is much smaller than 5% one would conclude that no significant influence of lateral heterogeneities in the mantle on earth tides could be measured by a gravimeter. However, Wang & Zschau (1985) suggested that the lateral heterogeneities might show dispersion and could be considerably stronger at tidal periods than at seismic periods, provided they are partly due to temperature differences.

The influences of heterogeneity as well as of inelasticity, the earth's rotation and elliptical stratification are some of the problems which are just starting to be considered. In order to separate the different contributions to earth tides, improvements in the instrumental techniques and in the theoretical models are required. Here we apply perturbation method based on the previous work of Molodenskij to determine the tidal deformation of a more realistic three-dimensional earth model. In this method a spherically symmetric, non-rotating, elastic and isotropic (SNREI) earth will be chosen as the starting earth model. All possible deviations such as the earth's rotation, inelasticity, anisotropy, lateral heterogeneities, and spherically asymmetric configurations including the geoid and interior boundary undulations as well as the elliptical stratification can be treated as parameter perturbations.

In Chapter 2 we use the linearized equations given by Dahlen (1972) to describe the incremental tidal deformations of a rotating, spherically asymmetrical, self-gravitating, pre-stressed earth. The definition of the hydrostatic pressure, which represents the principal part of the pre-stress, is generalized, so that it can be uniquely determined by the density and gravitational field. The boundary conditions are considered for a wide range of problems. The variation principle is formulated in Chapter 3. By making use of Green's formula, the solution of the disturbed equation system is directly determined by integrating the parameter perturbations multiplied by the known kernel functions.

In Chapter 4 all possible parameter perturbations are expanded in terms of spherical harmonics. By introducing the so-called auxiliary solutions, which in this case are the fundamental solutions of the boundary value problem of the undisturbed system, all the expansion coefficients of the sought solution components can be determined independently from each other by volume integrals, whose integrands depend only upon the undisturbed and auxiliary solutions and the model parameters. Furthermore, these volume integrals are separated to series of products of a one-dimensional radial integral and a two-dimensional surface integral, and the latter one will be solved analytically. The concrete formulas for the computational procedure are derived in Chapter 5.

Observational effects of earth tides as well as the station dependent tidal parameters are discussed in Chapter 6. Numerical results are presented in Chapter 7. First, we examine the numerical procedure by applying it to several special examples, which are solvable fully or partly analytically. In particular, we compute earth tides on a rotating, slightly elliptical and linearly elastic earth to provide a possible check for Wahr's model. Some important discrepancies are found and discussed. Finally, the influences of inelasticity and heterogeneity in the mantle are estimated for realistic models. A brief summary is offered in Chapter 8.

# Chapter 2. Differential Equations and Boundary Conditions

## 2.1 Initial equilibrium

The initial static equilibrium in the earth is commonly described by the following differential equations

$$(2\text{-}1)\qquad \begin{cases} \nabla\cdot\mathbf{\Sigma} + \rho\nabla V = 0\,, \\ \nabla^2 V = -4\pi G\rho + 2\Omega^2 \end{cases}$$

where $\mathbf{\Sigma}$ is the pre-stress tensor, $\rho$ the density distribution, G the gravitational constant, $\mathbf{\Omega}$ the angular velocity of the earth's rotation and

$$(2\text{-}2)\qquad V = \Phi + \Psi$$

is the gravity potential, consisting of the gravitational part

$$(2\text{-}3)\qquad \Phi = G\int_{\upsilon} \frac{\rho\, d^3\mathbf{r}^{o}}{|\mathbf{r} - \mathbf{r}^{o}|}$$

and the centrifugal part due to the earth's rotation

$$(2\text{-}4)\qquad \Psi = \frac{1}{2}\left(\,\Omega^2 r^2 - (\,\mathbf{\Omega}\cdot\mathbf{r}\,)^2\,\right).$$

The pre-stress tensor $\mathbf{\Sigma}$ is usually expressed as

$$(2\text{-}5)\qquad \mathbf{\Sigma} = -P\mathbf{I} + \mathbf{T}$$

where $\mathbf{I}$ is the unit tensor, $\mathbf{T}$ called the pre-stress deviator, and

$$(2\text{-}6)'\qquad P = -\frac{1}{3}\,\mathrm{tr}\,(\,\mathbf{\Sigma}\,)$$

the negative average of the three principal pre-stresses. In the hydrostatic case, the pre-stress deviator $\mathbf{T}$ vanishes and the parameter P is called the hydrostatic pressure satisfying the differential equation

$$(2\text{-}6)'' \qquad \nabla P = \rho \nabla \mathbf{V}.$$

In the general case, the pre-stress tensor $\mathbf{\Sigma}$ cannot be determined uniquely from the density and the potential (Dahlen, 1972). Furthermore, the parameter P is not a continuous function if the pre-stress shows discontinuities through a boundary. This discontinuity leads to difficulties in description of the boundary conditions for the tidal motion. However, if the earth's interior does not significantly deviate from the hydrostatic state, the pre-stress deviator can be treated as a small perturbation of first order. In this case the parameter P can be still uniquely determined from the density and the potential by omitting the rotational part of the vector $\rho \nabla \mathbf{V}$ in the right hand of the equation (2-6)'', or by the differential equation

$$(2\text{-}6) \qquad \nabla^2 P = \nabla \cdot ( \rho \nabla \mathbf{V} ).$$

The parameter P defined by the equation (2-6), instead of (2-6)', will be called in the following the generalized hydrostatic pressure. For its unique determination we additionally require that it is continuous through any interface in the earth's interior and vanishes at the earth's surface (if no air pressure is considered). In the case of the hydrostatic equilibrium, both the equations (2-6) and (2-6)', reduce to the identical definition of the hydrostatical pressure. In the general case it corresponds the isostatic pressure. By substituting the equation (2-5) and considering the equation (2-6), we find that the pre-stress deviator $\mathbf{T}$ satisfies the following differential equations

$$(2\text{-}7) \qquad \begin{cases} \nabla \times ( \nabla \cdot \mathbf{T} ) = \nabla \times ( \rho \nabla \mathbf{V} ) \\ \nabla \cdot ( \nabla \cdot \mathbf{T} ) = 0 . \end{cases}$$

As has been pointed out above, a unique determination of the pre-stress deviator $\mathbf{T}$ is impossible without additional information about, for example, the tectonic state in the earth's interior. However, in the earth's mantle it can be supposed to be much smaller than the corresponding hydrostatic pressure.

## 2.2 Differential equations for the incremental tidal deformations

The differential equations for the incremental tidal deformations superimposed on the initial equilibrium have been given by Dahlen (1972). Using the

Lagrangian reference frame, they are expressed as

$$
(2\text{-}8)\quad \begin{cases} \nabla\cdot\boldsymbol{\sigma} + \nabla(\mathbf{u}\cdot\nabla P) - (\nabla\cdot\mathbf{u})\nabla P - (\mathbf{u}\cdot\nabla)\nabla P + \rho\{\nabla\varphi + (\mathbf{u}\cdot\nabla)\nabla V\} \\ \quad - \rho\{\ddot{\mathbf{u}} + 2\boldsymbol{\Omega}\times\dot{\mathbf{u}}\} = \mathbf{0}, \\ \nabla^2\varphi - 4\pi G\nabla\cdot(\rho\mathbf{u}) = 0 \end{cases}
$$

where $\mathbf{u}$ is the incremental displacement vector, the dots over $\mathbf{u}$ denote the temporal derivatives, $\varphi$ is the incremental potential, and $\boldsymbol{\sigma}$ the incremental pseudo-stress tensor. Differing from the expressions given by Dahlen, in (2-8) the influence of the generalized hydrostatic pressure on the incremental pseudo-stress has been treated separately from the influence of the pre-stress deviator. The components of the incremental pseudo-stress tensor, $\sigma_{ij}$, in a Cartesian reference frame with its origin at the earth's centre of mass are given by

$$
(2\text{-}9)\quad \sigma_{ij} = \Lambda_{ijkl}\cdot u_{l,k}
$$

where $u_{i,j}$ ($i, j = 1, 2, 3$) denote space derivatives $\partial u_i/\partial x_j$, and Einstein's summation convention concerning repeated dummy subscripts is used. $\Lambda_{ijkl}$ ($i, j, k, l = 1, 2, 3$) are the components of the fourth order tensor $\boldsymbol{\Lambda}$ which describes the elastic coefficients for incremental deformations superimposed on the pre-stress deviator:

$$
(2\text{-}10)\quad \Lambda_{ijkl} = C_{ijkl} + \frac{1}{2}(T_{ij}\delta_{kl} + T_{kl}\delta_{ij} + T_{ik}\delta_{jl} - T_{jk}\delta_{il} - T_{il}\delta_{jk} + T_{jl}\delta_{ik})
$$

where $\delta_{ij}$ is Kronecker's substitution symbol, $T_{ij}$ the components of the pre-stress deviator, defined by (2-5), and $C_{ijkl}$ the 21 linear isentropic elastic coefficients.

## 2.3 Starting earth model SNREI

A special case is to consider a spherically symmetric, non-rotating, elastic and isotropic earth (SNREI). For static deformations, equation (2-8) reduces to

$$
(2\text{-}11)\quad \begin{cases} \rho^o\{\nabla(\varphi^o + u_r^o\cdot dV^o/dr) - (\nabla\cdot\mathbf{u}^o)dV^o/dr\,\mathbf{e}_r\} + \nabla\cdot\boldsymbol{\sigma}^o = \mathbf{0}, \\ \nabla^2\varphi^o - 4\pi G\nabla\cdot(\rho^o\mathbf{u}^o) = 0 \end{cases}
$$

where the spherical coordinate system is used, $\mathbf{e}_r$ denotes the unit vector in the radial direction, and the superscript $^o$ is used to distinguish the SNREI-parameters from the general ones in equation (2-8). Since in this case the pre-stress deviator vanishes

$$\text{(2-12)} \qquad \mathbf{T}^o = \mathbf{0},$$

the coefficient tensor $\Lambda_{ijkl}$ is simplified to

$$\text{(2-13)} \qquad \Lambda^o_{ijkl} = C^o_{ijkl} = \lambda^o \delta_{ij}\delta_{kl} + \mu^o(\delta_{ik}\delta_{jl} + \delta_{il}\delta_{jk})$$

and the pseudo-stress reduces to the familiar isotropic stress with components

$$\text{(2-14)} \qquad \sigma_{ij} = \lambda^o u_{k,k}\delta_{ij} + \mu^o(u_{i,j} + u_{j,i}).$$

## 2.4 Boundary conditions

There are three sets of boundary conditions for the incremental tidal deformations superimposed on the initial equilibrium. They have been deduced by Dahlen (1972). Since the influence of the generalized hydrostatic pressure has been excluded in the incremental pseudo-stress, the third boundary condition given here is formulated in another way differently from that given by Dahlen.

(a) Welded boundary (solid to solid)

$$\text{(2-15)} \qquad [\,\mathbf{u}\,]^+_- = 0$$

$$\text{(2-16)} \qquad \mathbf{n}\cdot[\,\nabla\varphi - 4\pi G\rho\mathbf{u}\,]^+_- = 0$$

$$\text{(2-17)'} \qquad [\,\mathbf{n}\cdot\boldsymbol{\sigma} - P\{(\nabla\cdot\mathbf{u})\mathbf{n} - (\nabla\mathbf{u})\cdot\mathbf{n}\}\,]^+_- = 0$$

where $[\,f\,]^+_- = f(\mathbf{r}^+) - f(\mathbf{r}^-)$ denotes the increment of function f through a boundary and $\mathbf{n}$ the unit normal of the boundary.

If a function is continuous through a boundary, its tangential derivatives along the boundary, if they exist, are also continuous through the boundary. Let us define the surface Nabla operator

(2-18) $\nabla_s = \nabla - \mathbf{n}(\mathbf{n}\cdot\nabla)$,

with the help of the continuity of the displacement $\mathbf{u}$, it can be deduced that

(2-19) $(\nabla\cdot\mathbf{u})\mathbf{n} - (\nabla\mathbf{u})\cdot\mathbf{n}$

$= (\nabla\cdot\mathbf{u})\mathbf{n} - \{\nabla_s\mathbf{u} + \mathbf{n}(\mathbf{n}\cdot\nabla)\mathbf{u}\}\cdot\mathbf{n}$

$= (\nabla\cdot\mathbf{u})\mathbf{n} - (\nabla_s\mathbf{u})\cdot\mathbf{n} - (\partial u_n/\partial n)\mathbf{n}$

$= (\nabla_s\cdot\mathbf{u})\mathbf{n} - (\nabla_s\mathbf{u})\cdot\mathbf{n}$

is continuous through a welded boundary. Using the definition (2-6) for the parameter P, the boundary condition (2-17)' simply becomes

(2-17) $[\,\mathbf{n}\cdot\boldsymbol{\sigma}\,]_-^+ = 0.$

(b) Frictionless boundary (fluid to fluid or fluid to solid)

(2-20) $[\,\mathbf{n}\cdot\mathbf{u}\,]_-^+ = 0,$

(2-16) $\mathbf{n}\cdot[\,\nabla\cdot\varphi - 4\pi G\rho\mathbf{u}\,]_-^+ = 0,$

(2-21)' $[\,\mathbf{n}\cdot\boldsymbol{\sigma} + P\{(\nabla\mathbf{u})\cdot\mathbf{n} - (\nabla_s\mathbf{u})\cdot\mathbf{n} - (\nabla\cdot\mathbf{u})\mathbf{n}\} + \nabla_s\cdot(P\mathbf{u})\mathbf{n}\,]_-^+ = 0.$

With arguments similar to those used in (a), (2-21)' is reduced to

(2-21) $[\,\mathbf{n}\cdot\boldsymbol{\sigma} + (\nabla_s P\cdot\mathbf{u})\mathbf{n}\,]_-^+ = 0.$

If the frictionless boundary is isobaric or the surface gradient $\nabla_s P$ to be small and negligible, the condition (2-21) becomes identical to (2-17).

(c) Summary

All boundary conditions can now be summarized as follows

$$\text{(2-22)}\quad \begin{cases} [\,\mathbf{u}\,]_-^+ = 0, & \text{welded;} \\ [\,\mathbf{n}\cdot\mathbf{u}\,]_-^+ = 0, & \text{frictionless.} \end{cases}$$

$$\text{(2-23)}\quad \mathbf{n}\cdot[\,\nabla\varphi - 4\pi G\rho\mathbf{u}\,]_-^+ = 0$$

$$\text{(2-24)}\quad [\,\mathbf{n}\cdot\boldsymbol{\sigma}\,]_-^+ = 0, \qquad \text{welded, or frictionless and isobaric.}$$

2.5 Surface conditions

We suppose that the earth's surface bears a time dependent surface dragging force $\mathbf{b}$. Its vertical component may be, for example, tidal ocean loading, or atmospheric pressure, or post-glacial unloading. Its tangential component may describe the frictional force at the ocean bottom due to tidal flow, or at the continental surface due to wind. The surface condition allowing for the surface dragging force $\mathbf{b}$ is commonly written as follows:

$$\text{(2-25)}\quad \mathbf{n}\cdot\boldsymbol{\sigma}\,|_{\partial\mathbf{v}} = \mathbf{b}$$

The other surface condition involving the potential is well known to be

$$\text{(2-26)}\quad \mathbf{n}\cdot[\,\nabla\varphi - 4\pi G\rho\mathbf{u}\,]_{\partial\mathbf{v}^-} = \mathbf{n}\cdot\nabla\varphi\,|_{\partial\mathbf{v}^+}\,.$$

## Chapter 3. Perturbation Method

In this chapter we apply perturbation theory to formulate influences of small irregular perturbations in the earth's parameters such as mantle heterogeneities on earth tides. The essential concept is based on the previous works of Molodenskij (1977, 1980; also see Molodenskij & Kramer, 1980). The variation principle and the Green's formula allow us to decouple the differential equations of the disturbed system and solve it by integrals which only depend on the model parameters and the fundamental solutions of the undisturbed system. Our improvements of this method are that we include perturbations of model parameters more completely and consider the effects not only in the gravity tides but also in all other tidal observables.

### 3.1 Definitions

We rewrite the equation of motion (2-8) for the disturbed system symbolically as follows

(3-1) $$\mathfrak{L}(\mathbf{y}) = 0,$$

defined for any $\mathbf{r} \in \mathfrak{v}$, where $\mathfrak{v}$ is the disturbed reference configuration, except for its surface $\partial\mathfrak{v}$ and interfaces $\partial\mathfrak{v}_i$ ($i = 1, 2, \cdots$). $\mathfrak{L}$ denotes the 4-dimensional differential vector operator depending on the disturbed earth's parameters $\rho$, V, P, $\mathbf{\Omega}$, $\mathbf{\Lambda}$, and the tidal frequency $\omega$.

(3-2) $$\mathbf{y} \equiv \{ \mathbf{u}, \varphi/(4\pi G) \}$$

defines the 4-dimensional solution vector. Analogously, equation (2-11) is rewritten as

(3-3) $$\mathfrak{L}^{o}(\mathbf{y}^{o}) = 0,$$

defined for any $\mathbf{r} \in \mathfrak{v}^{o}$, where $\mathfrak{v}^{o}$ is the undisturbed reference configuration, except for its surface $\partial\mathfrak{v}^{o}$ and interfaces $\partial\mathfrak{v}^{o}_i$ ($i = 1, 2, \cdots$). $\mathfrak{L}^{o}$ depends only upon the undisturbed earth's parameters $\rho^{o}$, $V^{o}$, $P^{o}$, and $\mathbf{\Lambda}^{o}$.

(3-4) $$\mathbf{y}^{o} \equiv \{ \mathbf{u}^{o}, \varphi^{o}/(4\pi G) \}$$

denotes the undisturbed solution vector. We introduce the following variations

$$\text{(3-5)} \quad \begin{cases} \delta\mathbf{S} = \mathbf{S} - \mathbf{S}^o \\ \delta\mathbf{y} = \mathbf{y} - \mathbf{y}^o . \end{cases}$$

In each intersect region $\mathbf{v}_i \cap \mathbf{v}_i^o$ (i = 1, 2, ···) they are defined uniquely, where $\mathbf{v}_i$ is the continuous region between interfaces $\partial\mathbf{v}_i$ and $\partial\mathbf{v}_{i-1}$, and so on. In order to limit all the variations to be small quantities of first order of the parameter perturbations, in the neighbourhood of $\partial\mathbf{v}_i^o$, that means in $\mathbf{v}_i \setminus \mathbf{v}_i^o$ and $\mathbf{v}_i^o \setminus \mathbf{v}_i$, the variations of all physical quantities are redefined to be smooth on either side of the undisturbed boundary by a linear extrapolation, for example,

$$\text{(3-6)} \quad \delta f(\mathbf{r}) = f(\mathbf{r}) - \{ f^o(\mathbf{r}^-)|_{\partial\mathbf{v}^o} + (\mathbf{r} - \mathbf{r}|_{\partial\mathbf{v}^o})\cdot\nabla f^o(\mathbf{r}^-)|_{\partial\mathbf{v}^o} \}$$

should be understood as the variation of the function f for any $\mathbf{r} \in \mathbf{v}_i \setminus \mathbf{v}_i^o$ and

$$\text{(3-7)} \quad \delta f(\mathbf{r}) = \{ f(\mathbf{r}^-)|_{\partial\mathbf{v}} + (\mathbf{r} - \mathbf{r}|_{\partial\mathbf{v}})\cdot\nabla f^o(\mathbf{r}) \} - f^o(\mathbf{r})$$

for any $\mathbf{r} \in \mathbf{v}_i^o \setminus \mathbf{v}_i$, where the boundary undulation

$$\text{(3-8)} \quad \delta h = \mathbf{n}^o \cdot ( \mathbf{r}|_{\partial\mathbf{v}} - \mathbf{r}|_{\partial\mathbf{v}^o} )$$

as well as its surface gradient $\nabla_s \delta h$ have been supposed to be perturbation quantities of first order.

3.2 Variation principle

We consider the variation of the disturbed system (3-1) to first order

$$\text{(3-9)} \quad \delta\mathbf{S}(\mathbf{y}^o) + \mathbf{S}^o(\delta\mathbf{y}) = 0$$

and suppose that the so-called auxiliary solution vectors

$$\text{(3-10)} \quad \mathbf{y}^j \equiv \{ \mathbf{u}^j, \varphi^j/(4\pi G) \}, \qquad (j = 1, 2, 3, 4)$$

satisfy the undisturbed system (3-3)

$$(3\text{-}11)\qquad \mathcal{L}^{o}(\mathbf{y}^{j}) = 0, \qquad\qquad j = 1,\ 2,\ 3,\ 4.$$

We now obtain an integral identity by integrating $\{\ \mathbf{y}^{j}\cdot(3\text{-}9) - \delta\mathbf{y}\cdot(3\text{-}11)\ \}$ over volume $\mathcal{V}^{o}$:

$$(3\text{-}12)\qquad \int_{\mathcal{V}^{o}} \{\ \mathbf{y}^{j}\cdot\delta\mathcal{L}(\mathbf{y}^{o})\ \}\, d^{3}\mathbf{r} + I^{j} = 0$$

where

$$(3\text{-}13)\qquad I^{j} = \int_{\mathcal{V}^{o}} \{\ \mathbf{y}^{j}\cdot\mathcal{L}^{o}(\delta\mathbf{y}) - \delta\mathbf{y}\cdot\mathcal{L}^{o}(\mathbf{y}^{j})\ \}\, d^{3}\mathbf{r}.$$

Since the differential operator $\mathcal{L}^{o}$ is self-conjugate, by means of Green's formula the integral (3-13) is equal to the sum of surface and interface ( if there are any interior interfaces) integrals. As will be shown below, these surface integrals may be expressed in terms of the sought variations of the solution (displacements and potential) at the earth's surface. By integrating by parts one may reduce equation (3-13) to

$$(3\text{-}14)\qquad I^{j} = \sum_{i}\int_{\partial\mathcal{V}^{o}_{i}} \mathbf{n}^{o}\cdot\Big[\ \boldsymbol{\sigma}^{j}\cdot\delta\mathbf{u} + \frac{\delta\varphi}{4\pi G}(\nabla\varphi^{j} - 4\pi G\rho^{o}\mathbf{u}^{j})$$
$$- \delta_{\mathbf{u}}\boldsymbol{\sigma}\cdot\mathbf{u}^{j} - \frac{\varphi^{j}}{4\pi G}(\nabla\delta\varphi - 4\pi G\rho^{o}\delta\mathbf{u})\ \Big]^{+}_{-} d^{2}\mathbf{r}$$
$$- \int_{\partial\mathcal{V}^{o}} \mathbf{n}^{o}\cdot\Big[\ \boldsymbol{\sigma}^{j}\cdot\delta\mathbf{u} + \frac{\delta\varphi}{4\pi G}(\nabla\varphi^{j} - 4\pi G\rho^{o}\mathbf{u}^{j})$$
$$- \delta_{\mathbf{u}}\boldsymbol{\sigma}\cdot\mathbf{u}^{o} - \frac{\varphi^{j}}{4\pi G}(\nabla\delta\varphi - 4\pi G\rho^{o}\delta\mathbf{u})\ \Big]^{-} d^{2}\mathbf{r}$$

where

$$(3\text{-}15)\qquad \boldsymbol{\sigma}^{j} = \lambda^{o}(\nabla\cdot\mathbf{u}^{j})\mathbf{I} + \mu^{o}[\nabla\mathbf{u}^{j} + (\nabla\mathbf{u}^{j})^{T}]$$

$$(3\text{-}16)\qquad \delta_{\mathbf{u}}\boldsymbol{\sigma} = \lambda^{o}(\nabla\cdot\delta\mathbf{u})\mathbf{I} + \mu^{o}[\nabla\delta\mathbf{u} + (\nabla\delta\mathbf{u})^{T}]$$

and $\mathbf{I}$ is the unit tensor of second order. Using the variations of boundary conditions as given below, the sum of the interface integrals in (3-14) may be expressed in terms which are independent of the solution variations. The sought solution variations at the earth's surface will be completely determined by substituting (3-14) into (3-12).

## 3.3 Variation of the boundary conditions

### (a) Welded boundary

Unless otherwise stated, all the variations in the present section are related to the undisturbed boundaries. Using the results in 2.4 and the definitions (3-6) and (3-7) it follows that

(3-17) $[\mathbf{n}^o \cdot \boldsymbol{\sigma}^j \cdot \delta\mathbf{u}]_-^+ = -\delta h[\mathbf{n}^o \cdot \boldsymbol{\sigma}^j \cdot \partial\mathbf{u}^o/\partial n^o]_-^+$

(3-18) $[\delta\varphi]_-^+ = -\delta h[\partial\varphi^o/\partial n^o]_-^+$

(3-19) $\nabla\varphi = \nabla\varphi + \delta h \cdot \partial\nabla\varphi/\partial n + \nabla\delta\varphi$

(3-20) $\rho = \rho^o + \delta h \cdot \partial\rho^o/\partial n^o + \delta\rho$

(3-21) $\mathbf{u} = \mathbf{u}^o + \delta h \cdot \partial\mathbf{u}^o/\partial n^o + \delta\mathbf{u}$

(3-22) $\boldsymbol{\sigma} = \boldsymbol{\sigma}^o + \delta h \cdot \partial\boldsymbol{\sigma}^o/\partial n^o + \delta_{\boldsymbol{\Lambda}}\boldsymbol{\sigma} + \delta_{\mathbf{u}}\boldsymbol{\sigma}$

where

(3-23) $\delta_{\boldsymbol{\Lambda}}\boldsymbol{\sigma} = \delta\boldsymbol{\Lambda} : \nabla\mathbf{u}^o$

is the variation of the pseudo-stress tensor due to variation in the coefficient tensor $\boldsymbol{\Lambda}$. The colon here denotes in our convention the scalar product of two tensors

$$\delta\boldsymbol{\Lambda} : \nabla\mathbf{u}^o = \delta\Lambda_{ijkl} \cdot u_{k,l} \,.$$

The left-hand terms in equations (3-19) - (3-22) are related to the disturbed boundary and the right-hand ones to the undisturbed boundary, respectively. Furthermore, the normal of the disturbed boundary may be expressed in terms of the normal of the corresponding undisturbed boundary and the surface gradient of the boundary undulation

(3-24) $$\mathbf{n} = \mathbf{n}^{o} - \nabla_{s}\delta h$$

Then with help of equations (3-19) - (3-24) one obtains

(3-25) $$\mathbf{n}^{o}\cdot[\ \nabla\delta\varphi - 4\pi G\rho^{o}\delta\mathbf{u}\ ]_{-}^{+}$$

$$= -\delta h[\ \partial(\partial\varphi^{o}/\partial n^{o} - 4\pi G\rho^{o}\mathbf{n}^{o}\cdot\mathbf{u}^{o})/n^{o}\ ]_{-}^{+}$$

$$+ 4\pi G[\ \delta\rho\,\mathbf{n}^{o}\cdot\mathbf{u}^{o}\ ]_{-}^{+} - 4\pi G\nabla_{s}\delta h\cdot[\ \rho^{o}\mathbf{u}^{o}\ ]_{-}^{+}$$

and

(3-26) $$\mathbf{n}^{o}\cdot[\ \delta_{\mathbf{u}}\boldsymbol{\sigma}\cdot\mathbf{u}^{j}\ ]_{-}^{+}$$

$$= -\mathbf{n}^{o}\cdot[\ \delta_{\Lambda}\boldsymbol{\sigma}\cdot\mathbf{u}^{j}\ ]_{-}^{+} + \nabla_{s}\delta h\cdot[\ \boldsymbol{\sigma}^{o}\cdot\mathbf{u}^{j}\ ]_{-}^{+} - \delta h\,\mathbf{n}^{o}\cdot[\ \partial\boldsymbol{\sigma}^{o}/\partial n^{o}\cdot\mathbf{u}^{j}\ ]_{-}^{+}$$

(b) Frictionless isobaric boundary

Taking into account that at a frictionless and isobaric boundary the surface dragging force $\mathbf{n}^{o}\cdot\boldsymbol{\sigma}^{o}$ as well as $\mathbf{n}^{o}\cdot\boldsymbol{\sigma}^{j}$ has no tangential component. Equations (3-17), (3-18), (3-25) and (3-26) for case (a) still hold.

(c) Earth's surface

Taking the variation of the surface conditions (2-25) and (2-26) and noting that the surface dragging force is the same for both the disturbed and the undisturbed system, the following relations hold

(3-27) $$\mathbf{n}^{o}\cdot\delta_{\mathbf{u}}\boldsymbol{\sigma} = \nabla_{s}\delta h\cdot\boldsymbol{\sigma}^{o} - \mathbf{n}^{o}\cdot\{\ \delta h\,\partial\boldsymbol{\sigma}^{o}/\partial n^{o} + \delta_{\Lambda}\boldsymbol{\sigma}^{o}\}$$

$$\text{(3-28)} \quad \mathbf{n}^o\cdot[\,\nabla\delta\varphi - 4\pi G\rho^o\delta\mathbf{u}\,]$$

$$= \mathbf{n}^o\cdot\{\,\nabla\delta\varphi|^+ + 4\pi G\delta\rho\,\mathbf{u}^o + \delta h[\,\partial\nabla\varphi^o/\partial n^o|^+$$

$$- \partial(\,\nabla\varphi^o - 4\pi G\rho^o\mathbf{u}^o\,)/\partial n^o\,]\,\} - 4\pi G\rho^o\mathbf{u}^o\cdot\nabla_s\,\delta h$$

where all the quantities are related to the inside of the undisturbed surface, unless specially denoted by $|^+$.

### 3.4 Perturbation method

After substitution of equations (3-17), (3-18), (3-27) and (3-28) into (3-14), one may obtain a more detailled formulation of the perturbation method than (3-12):

$$\text{(3-29)} \quad \int_{\partial\mathcal{V}^o} \mathbf{n}^o\cdot\left\{\boldsymbol{\sigma}^j\cdot\delta\mathbf{u}|^- + \frac{\delta\varphi}{4\pi G}(\,\nabla\varphi^j - 4\pi G\rho^o\mathbf{u}^j)|^- - \frac{\varphi^j}{4\pi G}\nabla\delta\varphi|^+\right\}d^2\mathbf{r}$$

$$= \int_{\mathcal{V}^o}\left\{\,\mathbf{y}^j\cdot\delta\mathbf{\mathfrak{B}}(\mathbf{y}^o)\,\right\}d^3\mathbf{r}$$

$$+ \sum_i \int_{\partial\mathcal{V}_i^o} \mathbf{n}^o\cdot\left\{\,\delta_{\boldsymbol{\Lambda}}\boldsymbol{\sigma}\cdot\mathbf{u}^j - \delta\rho\,\varphi^j\mathbf{u}^o\,\right\}d^2\mathbf{r} \qquad (*)$$

$$- \int_{\partial\mathcal{V}^o} \mathbf{n}^o\cdot\left\{\,\delta_{\boldsymbol{\Lambda}}\boldsymbol{\sigma}\cdot\mathbf{u}^j - \delta\rho\,\varphi^j\mathbf{u}^o\,\right\}d^2\mathbf{r} \qquad (**)$$

$$+ \sum_i \int_{\partial\mathcal{V}_i^o}\left\{\,\delta h\,\mathbf{n}^o\cdot[\,\partial\boldsymbol{\sigma}^o/\partial n^o\cdot\mathbf{u}^j - \boldsymbol{\sigma}^j\cdot\partial\mathbf{u}^o/\partial n^o\right.$$

$$+ \frac{\varphi^j}{4\pi G}\partial(\,\nabla\varphi^o - 4\pi G\rho^o\mathbf{u}^o\,)/\partial n^o - \frac{1}{4\pi G}\frac{\partial\varphi^o}{\partial n^o}(\,\nabla\varphi^j - 4\pi G\rho^o\mathbf{u}^j)\,]^+_-$$

$$\left. + \nabla_s\delta h\cdot[\,\rho^o\varphi^j\mathbf{u}^o - \boldsymbol{\sigma}^o\cdot\mathbf{u}^j\,]^+_-\right\}d^2\mathbf{r}$$

$$- \int_{\partial\mathcal{V}^o}\left\{\,\delta h\,\mathbf{n}^o\cdot[\,\partial\boldsymbol{\sigma}^o/\partial n^o\cdot\mathbf{u}^j + \frac{\varphi^j}{4\pi G}\partial(\,\nabla\varphi^o - 4\pi G\rho^o\mathbf{u}^o\,)/\partial n^o|^-\right.$$

$$\left. - \frac{\varphi^j}{4\pi G}\partial\nabla\varphi^o/\partial n^o|^+\,] + \nabla_s\delta h\cdot[\,\rho^o\varphi^j\mathbf{u}^o - \boldsymbol{\sigma}^o\cdot\mathbf{u}^j\,]^-\right\}d^2\mathbf{r}\,.$$

Converting the surface integrals (*) and (**) by means of Gauss theorem back into a volume integral and directly introducing the expression of $\mathbf{y}^j\cdot\delta\mathbf{\mathfrak{B}}$,

$$\mathbf{y}^j \cdot \delta\mathbf{S}(\mathbf{y}^o) = \mathbf{u}^j \cdot \Big\{ \mathbf{u}^o \cdot (\delta\rho\, \nabla\nabla V^o + \rho^o \nabla\nabla\delta V - \nabla\nabla\delta P) + \nabla(\mathbf{u}^o \cdot \nabla\delta P)$$
$$- (\nabla\cdot\mathbf{u}^o)\nabla\delta P - \rho^o[\, \ddot{\mathbf{u}}^o + 2\mathbf{\Omega}\times\dot{\mathbf{u}}^o\,] + \nabla\cdot\delta_{\mathbf{\Lambda}}\boldsymbol{\sigma} \Big\}$$
$$+ \varphi^j \Big\{ \nabla^2\varphi^o/(4\pi G) - \nabla\cdot(\delta\rho\,\mathbf{u}^o) \Big\},$$

(3-29) reduces to

(3-30) $$\int_{\partial\mathbf{v}^o} \mathbf{n}^o \cdot \Big\{ \boldsymbol{\sigma}^j \cdot \delta\mathbf{u}|^- + \frac{\delta\varphi}{4\pi G}(\nabla\varphi^j - 4\pi G\rho^o\mathbf{u}^j)|^- - \frac{\varphi^j}{4\pi G}\nabla\delta\varphi|^+ \Big\}\, d^2\mathbf{r}$$

$$= I^j(\Theta^o, \delta\Theta, \mathbf{y}^o, \mathbf{y}^j)$$

where $\Theta^o$ denotes the undisturbed earth's parameters ( $\rho^o$, $V^o$, $P^o$, $\mathbf{\Lambda}^o$ ) and $\delta\Theta$ their perturbations ( $\delta\rho$, $\delta V$, $\delta P$, $\delta\mathbf{\Lambda}$, $\delta h$, $\omega$, $\mathbf{\Omega}$ ). The integral $I^j$ is independent of the solution perturbation $\delta\mathbf{y} = ( \delta\mathbf{u}, \delta\varphi/4\pi G )$ and expressed by

(3-31) $$I^j = \int_{\mathbf{v}^o} f(\Theta^o, \delta\Theta, \mathbf{y}^o, \mathbf{y}^j)\, d^3\mathbf{r}$$

$$+ \sum_i \int_{\partial\mathbf{v}^o_i} F_i(\Theta^o, \delta\Theta, \mathbf{y}^o, \mathbf{y}^j)\, d^2\mathbf{r} + \int_{\partial\mathbf{v}^o} F_o(\Theta^o, \delta\Theta, \mathbf{y}^o, \mathbf{y}^j)\, d^2\mathbf{r}$$

where f, $F_i$ ( i = 1, 2, ... ) and $F_o$ are the so-called kernel functions, which are given by

(3-32) $$f = \mathbf{u}^j \cdot \Big\{ \mathbf{u}^o \cdot (\delta\rho\, \nabla\nabla V^o + \rho^o \nabla\nabla\delta V - \nabla\nabla\delta P) + \nabla(\mathbf{u}^o \cdot \nabla\delta P)$$
$$- (\nabla\cdot\mathbf{u}^o)\nabla\delta P - \rho^o[\, \ddot{\mathbf{u}}^o + 2\mathbf{\Omega}\times\dot{\mathbf{u}}^o\,] \Big\}$$
$$+ \delta\rho(\nabla\varphi^o \cdot \mathbf{u}^j + \nabla\varphi^j \cdot \mathbf{u}^o) - \delta_{\mathbf{\Lambda}}\boldsymbol{\sigma} : \nabla\mathbf{u}^j$$

(3-33) $$F_i = \delta h\, \mathbf{n}^o \cdot [\, \partial\boldsymbol{\sigma}^o/\partial n^o \cdot \mathbf{u}^j - \boldsymbol{\sigma}^j \cdot \partial\mathbf{u}^o/\partial n^o$$
$$+ \frac{\varphi^j}{4\pi G}\partial(\nabla\varphi^o - 4\pi G\rho^o\mathbf{u}^o)/\partial n^o - \frac{1}{4\pi G}\frac{\partial\varphi^o}{\partial n^o}(\nabla\varphi^j - 4\pi G\rho^o\mathbf{u}^j)\,]^+_-$$
$$+ \nabla_s\delta h \cdot [\, \rho^o\varphi^j\mathbf{u}^o - \boldsymbol{\sigma}^o \cdot \mathbf{u}^j\,]^+_-$$

$$(3\text{-}34)\qquad F_o = -\delta h\,\mathbf{n}^o\cdot[\,\partial\boldsymbol{\sigma}^o/\partial n^o\cdot\mathbf{u}^j + \frac{\varphi^j}{4\pi G}\partial(\nabla\varphi^o - 4\pi G\rho^o\mathbf{u}^o)/\partial n^o|^- - \frac{\varphi^j}{4\pi G}\partial\nabla\varphi^o/\partial n^o|^+\,] - \nabla_s\delta h\cdot[\,\rho^o\varphi^j\mathbf{u}^o - \boldsymbol{\sigma}^o\cdot\mathbf{u}^j\,]^-$$

For the auxiliary solutions $\mathbf{y}^j$ (j = 1, 2, 3, 4) chosen to satisfy the appropriate surface conditions, the sought solution perturbation $\delta\mathbf{y}$ at the earth's surface may now be determined. This will be carried out in the next chapters by expanding all the parameters and solutions and their perturbations in surface spherical harmonics.

## Chapter 4. Expansion of Parameter Perturbations in Terms of Spherical Harmonics

The real earth differs from geometrically simple models such as a symmetrical sphere with radially stratified properties. The most important perturbations are due to its surface not being a smooth sphere and to the presence of lateral heterogeneities of elastic properties and density. We assume that the large-scale tidal motion is hardly affected by local irregularities and consider the perturbations of long wavelengths only. It is then convenient to expand all the parameter perturbations in terms of spherical harmonics.

### 4.1 Normalized surface spherical harmonics

The fully normalized complex surface spherical harmonics are defined as follows:

(4-1) $$Y_{nm}(\vartheta, \lambda) = \left\{ \frac{(2n+1)(n-m)!}{4\pi (n+m)!} \right\}^{1/2} P_n^m(\cos\vartheta) \exp(i m \lambda)$$

for $n = 0, 1, 2, \cdots$ ; and $m = -n, \cdots, n$.

where $\vartheta$ is the geocentric co-latitude, $\lambda$ the geocentric longitude and $P_n^m(\cos\vartheta)$ the associated Legendre polynomials given by

(4-2) $$P_n^m(x) = \frac{(1-x^2)^{m/2}}{2^n n!} \frac{d^{n+m}}{dx^{n+m}} (x^2 - 1)^n$$

(4-3) $$P_n^{-m}(x) = (-1)^m \frac{(n-m)!}{(n+m)!} P_n^m(x)$$

The normalization condition is then

(4-4) $$\int_0^{2\pi} d\lambda \int_0^{\pi} d\vartheta \sin\vartheta \, Y_{nm}(\vartheta, \lambda) \cdot Y_{lk}^{*}(\vartheta, \lambda) = \delta_{nl} \cdot \delta_{mk}$$

where the asterisk denotes the complex conjugate.

To transform the integrals (3-30) into a convenient form for numerical evaluations, the important parameter deviations from the undisturbed SNREI system will be expanded in the following sections in terms of spherical harmonics.

### 4.2 Inertial force due to the earth's rotation

The influence of the frequency $\omega$ of the tide-generating force as well as of the angular velocity of the earth's rotation $\mathbf{\Omega}$ on tidal deformations is known to be small. These two parameters $\omega$ and $\mathbf{\Omega}$ may, therefore, be treated as small perturbations of first order. It will be further supposed that the angular velocity $\mathbf{\Omega}$ is uniform throughout the earth, time independent, and coincides with the polar axis $\mathbf{e}_z$. That means, no tidally induced nutation will be considered. The tidal frequency $\omega$ is a scalar constant or a single harmonic term of zero degree. The angular velocity $\mathbf{\Omega}$ can be expressed by the harmonic terms of first degree in the so-called spheroidal form:

$$\mathbf{\Omega} = \sqrt{4\pi/3}\,\Omega\left\{\mathbf{e}_r Y_{10} + \nabla_1 Y_{10}\right\} \tag{4-5}$$

where $\nabla_1 = \nabla - \mathbf{e}_r \partial/\partial r$ denotes the gradient along a spherical surface with its center in the origin of the coordinate system.

### 4.3 Lateral heterogeneities of elastic moduli

We model the lateral heterogeneities of the two isotropic Lamė coefficients with the method employed by Molodenskij (1980):

$$\text{(4-6)}\quad \left\{\begin{aligned} \delta\lambda &= \sum_l \sum_k \lambda_{lk}(r)\cdot Y_{lk}(\vartheta,\lambda) \\ \delta\mu &= \sum_l \sum_k \mu_{lk}(r)\cdot Y_{lk}(\vartheta,\lambda) \end{aligned}\right.$$

where the first index $l$ of the summation starts with $l = 1$ because the term of zero degree describes the radial heterogeneities. Even though here the same letter $\lambda$ is used for both the first Lamė coefficient and the geocentric longitude, they will not be confused.

### 4.4 Visco-elastic dispersion of the Lamė coefficients

According to the correspondence principle of Biot (1954) the Lamė coefficients in the visco-elastic case will be replaced by complex expressions. If the unrelaxed moduli have been chosen for the starting model SNREI and the visco-elastic effect is small ( a general supposition of perturbation method ), the dispersion of the Lamė coefficients due to the visco-elastic behaviour can then be treated as a perturbation. This perturbation can be included in (4-6) by taking into account that in this case the radially dependent coefficients are complex and the index $l$ will start with zero.

## 4.5 Hexagonal anisotropy

The hexagonal anisotropic Lamè coefficients in the spherical coordinates are defined by the following stress-strain relationship:

$$\sigma_{rr} = \lambda_{\perp} \Delta + 2\mu_{\perp} \varepsilon_{rr}$$

$$\sigma_{\vartheta\vartheta} = \lambda_{//} \Delta + 2\mu_{//} \varepsilon_{\vartheta\vartheta} + (\lambda_{\perp} - \lambda_{//}) \varepsilon_{rr}$$

$$\sigma_{\lambda\lambda} = \lambda_{//} \Delta + 2\mu_{//} \varepsilon_{\lambda\lambda} + (\lambda_{\perp} - \lambda_{//}) \varepsilon_{rr}$$

$$\sigma_{r\vartheta} = 2\mu_{\times} \varepsilon_{r\vartheta}$$

$$\sigma_{r\lambda} = 2\mu_{\times} \varepsilon_{r\lambda}$$

$$\sigma_{\vartheta\lambda} = 2\mu_{//} \varepsilon_{\vartheta\lambda}$$

where $\Delta$ denotes the volume expansion (dilatation):

$$\Delta = \varepsilon_{rr} + \varepsilon_{\vartheta\vartheta} + \varepsilon_{\lambda\lambda}.$$

The anisotropic perturbations $\delta f = \delta\lambda_{\perp}, \delta\lambda_{//}, \delta\mu_{\perp}, \delta\mu_{//}$ and $\delta\mu_{\times}$ take the same form:

$$\delta f = \sum_{l} \sum_{k} f_{lk}(r) \cdot Y_{lk}(\vartheta, \lambda) \tag{4-7}$$

## 4.6 Lateral heterogeneity of density distribution

Molodenskij & Kramer (1980) argued that the lateral heterogeneity of the density distribution in the earth's mantle is rather small and can be neglected. However, it is not known whether its influence on earth tides, particularly on tidal gravity, is negligible. We model the density perturbation as follows:

$$\delta\rho = \sum_{l} \sum_{k} \rho_{lk}(r) \cdot Y_{lk}(\vartheta, \lambda) \tag{4-8}$$

4.7 Boundary undulations

We suppose that all the boundary undulations as well as their tangential derivatives are small:

(4-9) $$\delta h = \sum_l \sum_k h_{lk}(r) \cdot Y_{lk}(\vartheta, \lambda),$$

(4-10) $$\nabla_s \delta h = \sum_l \sum_k h_{lk}(r) \cdot \nabla_1 Y_{lk}(\vartheta, \lambda).$$

4.8 Perturbation of the gravity field

The perturbation of the gravity field is not an independent quantity. Since the influence of the earth's rotation has been treated as a perturbation, the perturbation of the gravity field consists of the centrifugal part $\Psi$ (defined by (2-4)) and the gravitational part $\delta\Phi$ due to the laterally heterogeneous density distribution and the boundary undulations:

(4-11) $$\delta V = \Psi + \delta\Phi.$$

The expansion series of the centrifugal part in terms of the normalized spherical harmonics is

(4-12) $$\Psi = \sum_l \sum_k \Psi_{lk}(r) \cdot Y_{lk}(\vartheta, \lambda)$$

with $\Psi_{00} = \sqrt{4\pi/9} \cdot \Omega^2 r^2$, $\Psi_{20} = -\sqrt{4\pi/45} \cdot \Omega^2 r^2$, and $\Psi_{lk} = 0$ otherwise. The gravitational part is determined by

(4-13) $$\delta\Phi = \int_{\mathbf{U}^o} \frac{\delta\rho \, d^3\mathbf{r}^o}{|\mathbf{r} - \mathbf{r}^o|} + \sum_i G \int_{\partial \mathbf{U}_i^o} \frac{-\Delta\rho\,\delta h}{|\mathbf{r} - \mathbf{r}^o|} d^2\mathbf{r}^o$$

$$= \sum_l \sum_k \Phi_{lk}(r) \cdot Y_{lk}(\vartheta, \lambda)$$

where $-\Delta\rho\,\delta h = -[\rho^o]_-^+ \delta h$ is the so-called surface density, and the sum of interface integrals includes also that for the outer surface, where the density of the air is supposed to be zero. By expanding the function $1/|\mathbf{r} - \mathbf{r}^o|$ in terms of spherical harmonics, the coefficients $\Phi_{lk}$ can be expressed by

(4-14) $$\Phi_{lk} = \frac{4\pi G}{(2l+1)} \left\{ \left(\frac{1}{r}\right)^{l+1} \int_o^r \rho_{lk} r^{l+2} dr + r^l \int_r^a \rho_{lk} \left(\frac{1}{r}\right)^{l-1} dr \right.$$

$$\left. - \sum_i \Delta\rho^o r^i h_{lk} \left[ \left(\frac{r}{r^i}\right)^l H(r^i - r) + \left(\frac{r^i}{r}\right)^{l+1} H(r - r^i) \right] \right\}$$

where $r^i$ denotes the radius of the i-th interior boundary, and H is the step function defined by

(4-15) $$H(x) = \begin{cases} 0, & x \leq 0; \\ 1, & x > 0. \end{cases}$$

It should be noted that the perturbation $\delta V$ still is a continuous function in the same way as the potential V itself.

4.9 Perturbation of the generalized hydrostatic pressure

In the undisturbed spherically symmetric system, it holds that

$$\nabla P^o = \rho^o \nabla V^o$$

$$\nabla V^o = - g^o \mathbf{e}_r$$

$$\nabla^2 V^o = - 4\pi G \rho^o$$

$$\nabla \rho^o = \dot{\rho}^o \mathbf{e}_r$$

where $\dot{\rho}^o$ denotes the radial derivative $d\rho^o/dr$. Furthermore, the potential perturbation satisfies the following Poisson's equation

(4-16) $$\nabla^2 \delta V = - 4\pi G \delta\rho + 2\Omega^2$$

By varying equation (2-6) the corresponding Poisson's equation for the perturbation of the generalized hydrostatic pressure can be reduced to

(4-17) $$\nabla^2 \delta P = - g^o \partial\delta\rho/\partial r + \dot{\rho}^o \partial\delta V/\partial r + \rho^o (2\Omega^2 - 8\pi G \delta\rho)$$

$$= - 4\pi G \sum_l \sum_k Q_{lk}(r) \cdot Y_{lk}(\vartheta, \lambda)$$

and

$$Q_{lk} = 2\rho^o \rho_{lk} + \frac{1}{4\pi G}(g^o \partial\rho_{lk}/\partial r - \dot{\rho}^o \partial V_{lk}/\partial r) - \frac{2}{\sqrt{4\pi}\,G}\Omega^2 \rho^o \delta_{l,o}\delta_{k,o} \tag{4-18}$$

with $V_{lk} = \Psi_{lk} + \Phi_{lk}$. The associated boundary condition is

$$[\,\delta P\,]_-^+ = g^o \Delta\rho^o \delta h. \tag{4-19}$$

This boundary value problem can be solved analytically by potential theory. The solution is

$$\delta P = \sum_l \sum_k P_{lk}(r) \cdot Y_{lk}(\vartheta, \lambda) \tag{4-20}$$

$$P_{lk} = \frac{4\pi G}{(2l+1)} \left\{ \left(\frac{1}{r}\right)^{l+1} \int_o^r Q_{lk}\, r^{l+2} dr + r^l \int_r^a Q_{lk} \left(\frac{1}{r}\right)^{l-1} dr \right\} \tag{4-21}$$

$$- \sum_i g^o \Delta\rho^o h_{lk} \cdot \left(\frac{r}{r^i}\right)^l H(r^i - r)$$

$$- \frac{4\pi G}{(2l+1)}\, a \left(\frac{r}{a}\right)^l \int_o^a Q_{lk} \left(\frac{r}{a}\right)^{l+2} dr$$

where a is the earth's average radius. The first term in (4-21) is a special solution of Poisson's equation (4-17), the second and the last term are additional harmonics, introduced in order to satisfy the boundary condition (4-19) at all interfaces and the earth's surface.

We here again emphasize that a unique determination of the generalized hydrostatic pressure is only possible, when the new definition (2-6) is introduced.

### 4.10 Deviatory pre-stress

The deviatory pre-stress is the other quantity which depends upon density, boundary undulations, the gravity field and the generalized hydrostatic pressure. As stated before, it cannot be uniquely determined. However, in Chapter 2 it has been shown that this parameter occurs only together with the elastic moduli in a comparable position in equation (2-10). From the present knowledge of seismology we know that, on one hand, rocks cannot bear a shear stress greater than about 1 kbar. The seismically observed velocity heterogeneities, on the other hand, can reach up to a few per cent in the mantle. Furthermore, if the seismically observed heterogeneities are temperature induced, they will be much more pronounced in the tidal band than in the seismic band and can be intensified by a factor of 3-5, based on our model calculations (see Chapter 7), so that in the tidal band the lateral heterogeneities of the elastic moduli in the mantle can reach tens of kbar, much greater than the critical shear stress in rocks. So the influences of the deviatory part of the pre-stresses will be neglected below.

## Chapter 5. Determination of Auxiliary, Undisturbed, and Disturbed Solutions

In general, the disturbed solution includes both the spheroidal and toroidal modes with an infinite set of coefficients. As has been stated, they are coupled together subject to specific coupling rules. In the Wahr-Smith theory (see Wahr, 1979) of the modification of the earth tides due to the earth's rotation and elliptical stratification, the parameter perturbations are terms of spherical harmonics of second degree. Though Wahr truncates the infinite set of differential equations and variables by neglecting terms smaller than of the first order of the ellipticity, the truncated system consists in the general case of 22 coupled equations. He had to solve this system in two steps. In the first step, a super-truncated system of 10 ordinary differential equations with 6 coefficients of the spheroidal modes and 4 coefficients of the toroidal modes are integrated. The obtained solution is used in the second step as an approximation for determining the remaining unknowns by a perturbation method. Thus, in principle, Wahr's direct integration method takes the same truncation as the present perturbation method.

In the present perturbation method, the disturbed system of differential equations is truncated by the variation principle and converted to a finite system of integral equations by making use of Green's formula. In particular, this method introduces the so-called auxiliary solutions which, as will be shown, are the fundamental solutions of the undisturbed spherical system in connection with certain boundary value problems. By using these auxiliary solutions the system can be decoupled and all the coefficients of the truncated solution can be evaluated independently from each other.

In this chapter, we first formulate the undisturbed solution and the auxiliary solutions in the general form which also includes the toroidal modes. The generalization of the undisturbed solution is necessary if deformations caused by surface shear stresses, due to e.g. wind, tides or tidal friction on the continental shelfs are considered. The generalization of the auxiliary solutions is necessary if the toroidal effects in the tidal observables, especially in the tidal displacement and tilt due to asymmetric parameter distributions, are not negligible. In Section 5.3 we decouple the system of integral equations. Here the four coefficient groups of the solution variation are determined by independent integrals in which the four independent auxiliary solutions are chosen separately. Some problems in connection with determining the coefficients of degree 0 and 1 are discussed in Section 5.4. The integrands, often called kernel functions, are determined in Section 5.5 in the form convenient for the computational procedure. The formulation process is complicated. For simplicity, we demonstrate only an example for the influence of heterogeneity of the shear modulus. The complete formulae are given in Appendix C.

### 5.1 Undisturbed solutions

Unless otherwise stated, the time factor $\exp(i\omega t)$ for both the undisturbed and disturbed solution will be omitted below. The undisturbed (starting) solution vector $\mathbf{y}^o = ( \mathbf{u}^o, \varphi^o / 4\pi G )$ of degree $n_o$ and order $m_o$ is supposed to take the following form:

$$\text{(5-1)} \qquad \mathbf{u}^o = \left\{ H_{n_o}(r)\,\mathbf{e}_r + T_{n_o}(r)\,\nabla_1 + W_{n_o}(r)\,\mathbf{e}_r \times \nabla_1 \right\} Y_{n_o m_o}(\vartheta, \lambda)$$

$$\text{(5-2)} \qquad \varphi^o = R_{n_o}(r)\, Y_{n_o m_o}(\vartheta, \lambda).$$

The radially dependent coefficients H, T and R are known as the coefficient set for describing the spheroidal mode and in the case of body tides correspond to the Love numbers h, l and k, respectively. Coefficient W describes the toroidal mode. In the spherically symmetric system it only exists when a shear stress is exerted. In the perturbation theory, the undisturbed solutions are assumed to be known for the starting model SNREI.

## 5.2 Auxiliary solutions

The auxiliary solutions, first introduced by Molodenskij (1977), of degree n and order m are generalized by the following expressions:

$$\text{(5-3)} \qquad \mathbf{u}^j = \left\{ H_n^j(r)\,\mathbf{e}_r + T_n^j(r)\,\nabla_1 + W_n^j(r)\,\mathbf{e}_r \times \nabla_1 \right\} Y_{nm}^*(\vartheta, \lambda)$$

$$\text{(5-4)} \qquad \varphi^j = R_n^j(r) \cdot Y_{nm}^*(\vartheta, \lambda).$$

There are four independent auxiliary solutions, three for the spheroidal mode and one for the toroidal mode. All of them satisfy the undisturbed differential equations and can be evaluated by a procedure similar to the one used for determining the normal modes. The associated surface conditions will be chosen, so that the components of the disturbed solution vector at the earth's surface can be determined independently of each other.

## 5.3 Solution perturbation

The sought solution perturbations at the earth's surface which are, strictly speaking, related to the inside of the undisturbed surface, will be expressed in the expansion series of spherical harmonics:

$$\text{(5-5)} \qquad \delta\mathbf{u} = \sum_n \sum_m \left\{ H_n^m\,\mathbf{e}_r + T_n^m\,\nabla_1 + W_n^m\,\mathbf{e}_r \times \nabla_1 \right\} Y_{nm}(\vartheta, \lambda)$$

$$\text{(5-6)} \qquad \delta\varphi = \sum_n \sum_m R_n^m\, Y_{nm}(\vartheta, \lambda).$$

**The corresponding values related to the disturbed surface may be obtained by the linear extrapolation given in Section 3.1. For the potential perturbation $\delta\varphi$, it is useful to determine its value at a point related to the outside of the undisturbed surface. By using the variation of the boundary conditions as stated in Section 3.3, we obtain**

$$\begin{aligned}\delta\varphi|^{+} &= \delta\varphi|^{-} - \delta h[\partial\varphi^{o}/\partial r]_{-}^{+} \\ &= \delta\varphi|^{-} + 4\pi G\rho^{o} u_{r}^{o}\,\delta h \\ &= \sum_{n}\sum_{m} R_{n}^{m} Y_{nm}(\vartheta,\lambda) + 4\pi G\rho^{o} H_{n_o} Y_{n_o m_o}(\vartheta,\lambda)\sum_{l}\sum_{k} h_{lk} Y_{lk}(\vartheta,\lambda).\end{aligned}$$

Expanding the product of the two spherical harmonics leads to a standard form for $\delta\varphi|^{+}$:

$$\delta\varphi|^{+} = \sum_{n}\sum_{m}\left\{ R_{n}^{m} + 4\pi G\rho^{o} H_{n_o}\sum_{l}\sum_{k}\tilde{A}(n_o, m_o; n, m; l, k)\, h_{lk}\right\} Y_{nm}(\vartheta,\lambda) \tag{5-7}$$

where

$$\begin{aligned}\tilde{A}(n_o, m_o; n, m; l, k) &= \int_{o}^{2\pi} d\lambda \int_{o}^{\pi} d\vartheta \sin\vartheta\, Y_{n_o m_o} Y_{nm}^{*} Y_{lk} \\ &= (-1)^{m}\int_{o}^{2\pi} d\lambda \int_{o}^{\pi} d\vartheta \sin\vartheta\, Y_{n_o m_o} Y_{n,-m} Y_{lk} \\ &= (-1)^{m} A(n_o, m_o; n, -m; l, k)\end{aligned} \tag{5-8}$$

is the expansion coefficient which is determined in Appendix B. Since the potential perturbation $\delta\varphi$ outside the earth is a harmonic function, for $r > a$, it takes the form

$$\begin{aligned}\delta\varphi = \sum_{n}\sum_{m}&\left\{ R_{n}^{m} + 4\pi G\rho^{o} H_{n_o}\sum_{l}\sum_{k}\tilde{A}(n_o, m_o; n, m; l, k)\, h_{lk}\right\} \\ &\cdot\left(\frac{a}{r}\right)^{n+1} Y_{nm}(\vartheta,\lambda), \qquad (r > a).\end{aligned} \tag{5-9}$$

and

$$(5\text{-}10)\quad \frac{\partial\delta\varphi}{\partial r}\Big|_a^+ = -\sum_n\sum_m \frac{(n+1)}{a}\Big\{ R_n^m + 4\pi G\rho^o H_{n_o} \sum_l\sum_k \tilde{A}(n_o, m_o; n, m; l, k)\, h_{lk} \Big\}$$

$$\times\, Y_{nm}(\vartheta, \lambda).$$

In Section 3.4 we have deduced the integral identity:

$$(3\text{-}30)\quad \int_{\partial \mathbf{v}^o} \mathbf{n}^o \cdot \Big\{ \boldsymbol{\sigma}^j \cdot \delta\mathbf{u}|^- + \frac{\delta\varphi}{4\pi G}(\nabla\varphi^j - 4\pi G\rho^o \mathbf{u}^j)|^- - \frac{\varphi^j}{4\pi G}\nabla\delta\varphi|^+ \Big\}\, d^2\mathbf{r}$$

$$= I^j(\Theta^o, \delta\Theta, \mathbf{y}^o, \mathbf{y}^j)$$

where the integral $I^j$ is independent of the solution perturbation and given by equations (3-31) - (3-34). By substituting the expansion series of spherical harmonics (5-1) - (5-7) and (5-10) into (3-30) and replacing the undisturbed boundary normal, $\mathbf{n}^o$, by the unit radial vector, $\mathbf{e}_r$, we obtain for degree n and order m the following relation:

$$(5\text{-}11)\quad a^2 \Big\{ y_2^j H_n^m + \frac{1}{4\pi G} y_6^j R_n^m + n(n+1)[y_4^j T_n^m + y_8^j W_n^m] \Big\}$$

$$= \Gamma_{nm}^j(\Theta^o, \delta\Theta, \mathbf{y}^o, \mathbf{y}^j)$$

where

$$(5\text{-}12)\quad \Gamma_{nm}^j(\Theta^o, \delta\Theta, \mathbf{y}^o, \mathbf{y}^j)$$

$$= I_{nm}^j(\Theta^o, \delta\Theta, \mathbf{y}^o, \mathbf{y}^j) + \int_{\partial \mathbf{v}^o} \frac{\varphi^j}{4\pi G} \Big\{ \frac{\partial}{\partial r}\delta\varphi_{nm}\Big|^+ + \frac{(n+1)}{a}\delta\varphi_{nm}\Big|^- \Big\}\, d^2\mathbf{r}$$

$$= I_{nm}^j(\Theta^o, \delta\Theta, \mathbf{y}^o, \mathbf{y}^j) - (n+1)\, a\rho^o H_{n_o} R_n^j \sum_l\sum_k h_{lk}\tilde{A}(n_o, m_o; n, m, l, k)\Big|_{\partial \mathbf{v}^o}^-$$

Furthermore, in (5-11) the notations which are usually introduced in the boundary problem of the spherically symmetric system have been used:

$$(5\text{-}13)\quad y_2^j = \lambda^o\{ \dot{H}_n^j + 2H_n^j/r - n(n+1)T_n^j/r \} + 2\mu^o \dot{H}_n^j$$

(S-14) $$y_4^j = \mu^o \left\{ \dot{T}_n^j + H_n^j / r - T_n^j / r \right\}$$

(S-15) $$y_6^j = \dot{R}_n^j + (n+1) R_n^j / r - 4\pi G \rho^o H_n^j$$

(S-16) $$y_8^j = \mu^o \left\{ \dot{W}_n^j - W_n^j / r \right\}$$

and the following orthogonalities of spherical harmonics have been considered:

$$\int_o^{2\pi} d\lambda \int_o^{\pi} d\vartheta \sin\vartheta \, Y_{nm} Y_{nm}^* = 1$$

$$\int_o^{2\pi} d\lambda \int_o^{\pi} d\vartheta \sin\vartheta \, \nabla_1 Y_{nm} \cdot \nabla_1 Y_{nm}^* = n(n+1)$$

$$\int_o^{2\pi} d\lambda \int_o^{\pi} d\vartheta \frac{\partial}{\partial\vartheta} \left\{ Y_{nm} Y_{nm}^* \right\} = 0$$

For degree $n \geq 2$, let us choose the auxiliary solutions satisfying the following surface conditions:

(S-17) $$C^j \cdot \left\{ y_2^j, y_4^j, y_6^j, y_8^j \right\} = \begin{cases} \{ 1, 0, 0, 0 \}, & j = 1; \\ \{ 0, \frac{1}{n(n+1)}, 0, 0 \}, & j = 2; \\ \{ 0, 0, 4\pi G, 0 \}, & j = 3; \\ \{ 0, 0, 0, \frac{1}{n(n+1)} \}, & j = 4; \end{cases} \qquad n \geq 2.$$

where $C^j$ ( $j = 1, 2, 3, 4$ ) are non-zero constants. As the auxiliary solutions enter linearly into equation (S-11), these constants have no influence on the solution perturbation and can be chosen arbitrarily. Substituting the auxiliary solutions into (S-11) successively, the explicit expressions of the sought expansion coefficients of the solution perturbation can be determined:

$$\text{(5-18)} \qquad \Gamma^{j}_{nm}(\, \Theta^{o},\, \delta\Theta,\, \mathbf{y}^{o},\, \mathbf{y}^{j}\,)\frac{C^{j}}{a^{2}} = \begin{cases} H^{m}_{n}, & j = 1; \\ T^{m}_{n}, & j = 2; \\ R^{m}_{n}, & j = 3; \\ W^{m}_{n}, & j = 4; \end{cases} \qquad n \geq 2.$$

## 5.4 Some problems in determining the solutions for degrees n = 0 and n = 1

The auxiliary solutions for degrees $n \geq 2$ are uniquely determined by choosing the surface conditions (5-17) for the auxiliary solutions. For the degree $n = 0$ and $n = 1$, however, they should be considered specially.

In fact, coefficient $T^{o}_{o}$ and $W^{o}_{o}$ have no meaning because the spherical harmonic of zero degree is a constant and its derivative vanishes. The component of $\delta\varphi$ of degree $n = 0$ in the free space must vanish because of conservation of the total mass. Therefore, only the first auxiliary solution ($j = 1$) is needed for case $n = 0$ and will be used for determining the uniform volume variation of the earth.

The solutions for degree $n = 1$ depend upon selection of the reference coordinate frame. In this case the toroidal modes with coefficients $W^{m}_{n}$ ($n = 1$, $m = -1, 0, 1$) describe the tidal nutation of the earth. Since we are interested in the tidal deformations, these modes will be omitted below. Furthermore, in the reference frame with its origin at the centre of mass the potential perturbation $\delta\varphi$, outside the earth, just as in case $n = 0$, contains no components of degree $n = 1$. Therefore, only the first two auxiliary solutions ($j = 1$ and $j = 2$) will be needed for degree $n = 1$. Since the translational motion of the earth as a rigid body satisfies the undisturbed equations with the free surface conditions, the three surface conditions for the auxiliary solutions of the spheroidal mode of degree $n = 1$ must be dependent upon each other with the following relation (Farrell, 1972):

$$\text{(5-19)} \qquad y^{j}_{2} + 2y^{j}_{4} + \frac{g^{o}}{4\pi G}\, y^{j}_{6} = 0$$

The surface conditions (5-17) should thus be modified by putting additionally

$$\text{(5-20)} \qquad y^{j}_{6} = -\, 4\pi G/(\, C^{j} g^{o}\,), \qquad \text{for } j = 1,\ j = 2 \text{ and } n = 1.$$

In order to uniquely determine the auxiliary solutions for the case $n = 1$, an additional surface condition will be required. We shall formulate it by using the fact that the coordinates of the mass centre are zero in the reference frame with its origin at the centre of mass. To first order we have

$$\int_{\mathfrak{V}^o} (\rho^o + \rho^1)\,\mathbf{r}\,d^3\mathbf{r} + \int_{\partial \mathfrak{V}^o} \rho^o u_r^j\,\mathbf{r}\,d^2\mathbf{r} = 0 \tag{5-21}$$

where

$$\rho^1 = -\nabla\cdot(\rho^o \mathbf{u}^j) = -\nabla^2 \varphi^j / 4\pi G \tag{5-22}$$

is the density variation during deformation and for simplicity, it has been supposed that there are no interior interfaces with density discontinuity (In fact, any density distribution in the neighbourhood of the discontinuity in practice can be replaced by a continuous function having a sufficiently steep gradient). Since already

$$\int_{\mathfrak{V}^o} \rho^o \mathbf{r}\,d^3\mathbf{r} = 0, \tag{5-23}$$

by substituting (5-22) and (5-23) into (5-21) it follows

$$\int_{\mathfrak{V}^o} \mathbf{r}\,\nabla^2 \varphi^j\,d^3\mathbf{r} - 4\pi G \int_{\partial \mathfrak{V}^o} \rho^o u_r^j\,\mathbf{r}\,d^2\mathbf{r} = 0\,.$$

or

$$\int_{\mathfrak{V}^o} \nabla\cdot(\nabla\varphi^j \mathbf{r} - \varphi^j \mathbf{I})\,d^3\mathbf{r} - 4\pi G \int_{\partial \mathfrak{V}^o} \rho^o u_r^j\,\mathbf{r}\,d^2\mathbf{r} = 0 \tag{5-24}$$

Converting the volume integral to a surface integral by means of Gauss theorem and using the expanded expression of $\varphi^j$, this integral identity is reduced to

$$\int_{\partial \mathfrak{V}^o} \left\{ \dot{R}_n^j - R_n^j / r - 4\pi G \rho^o H_n^j \right\} Y_{nm}(\vartheta, \lambda)\cdot \mathbf{r}\,d^2\mathbf{r} = 0 \tag{5-25}$$

or

$$\int_{\partial \mathfrak{V}^o} \left\{ y_6^j - (n+2) R_n^j / r \right\} Y_{nm}(\vartheta, \lambda)\cdot \mathbf{r}\,d^2\mathbf{r} = 0. \tag{5-26}$$

Considering that the Cartesian components of the position vector $\mathbf{r}$ are spherical harmonics of first degree n = 1, the sought additional surface condition can then be deduced from (5-26):

(5-27) $\quad R_n^j = y_6^j\, a/3 = -4\pi G a/(3C^j g^o)$, for j = 1, j = 2 and n = 1.

To summarize for degree n = 0 and n = 1, we arrive at the following conclusions:

(a) Some coefficients may be determined directly by physical reasoning:

(5-28) $\quad T_n^m = 0,$ for n = 0;

(5-29) $\quad R_n^m = -4\pi G\rho^o H_{n_o} \sum_l \sum_k \tilde{A}(n_o, m_o; n, m; l, k)\, h_{lk}|_{\bar{\partial} \mathbf{v}^o}$, for n = 0, 1;

(5-30) $\quad W_n^m = 0,$ for n = 0, 1.

(b) Surface conditions (5-17) for degree n = 0 and 1 should be modified to

(5-31) $\quad C^j \cdot \{ y_2^j, y_4^j, y_6^j, y_8^j \} = \{ 1, 0, 0, 0 \},$ for j = 1 and n = 0;

(5-32) $$C^j \cdot \{ y_2^j, y_4^j, y_6^j, y_8^j \} = \begin{cases} \{ 1, 0, -4\pi G/g^o, 0 \}, & j = 1, \\ \{ 0, 1/2, -4\pi G/g^o, 0 \}, & j = 2, \end{cases} \quad \text{and } n = 1.$$

(c) Identity (5-18), consequently, should be modified by a correction term:

(5-33) $$\Gamma_{nm}^j(\Theta^o, \delta\Theta, \mathbf{y}^o, \mathbf{y}^j)\frac{C^j}{a^2} - \frac{4\pi G\rho^o H_{n_o}}{g^o} \sum_l \sum_k \tilde{A}(n_o, m_o; n, m; l, k)\, h_{lk}|_{\bar{\partial} \mathbf{v}^o}$$

$$= \begin{cases} H_n^m, & \text{for } j = 1, \\ T_n^m, & \text{for } j = 2, \end{cases} \quad \text{and } n = 1.$$

5.5 Evaluation of integrals $\Gamma_{nm}^j$

It has been shown that the tidal anomalies at the earth's surface can be evaluated by integrals $\Gamma_{nm}^j$ (or $I_{nm}^j$) whose integrands are expressed in terms of triple products of the parameter perturbations, the undisturbed and the auxiliary solutions, or their derivatives. Using the expansions in spherical harmonics the volume integrals may be separated into series of products of radial and surface integrals. The radial integrals will be carried out from the earth's centre to the surface and, in general, have to be calculated numerically, while the surface integrals, which are composed only of triple products of spherical harmonics or their derivatives, will be carried out over an unit sphere surface and can be solved analytically (s. Appendix B).

The procedure of separating the volume integrals into the radial and spherial surface integrals is complicated. We give the explicit formulae in Appendix C and demonstrate here an example with respect to heterogeneities of the Lamė coefficient $\delta\mu$:

$$(5\text{-}34)\qquad \Gamma^{j}_{nm}(\delta\mu) = -\iiint_{r\le a} \delta\mu\,[\,\nabla\mathbf{u}^{o} + (\nabla\mathbf{u}^{o})^{T}\,] : \nabla\mathbf{u}^{j}\,d\upsilon$$

$$= -\frac{1}{2}\iiint_{r\le a} \delta\mu\,[\,\nabla\mathbf{u}^{o} + (\nabla\mathbf{u}^{o})^{T}\,] : [\,\nabla\mathbf{u}^{j} + (\nabla\mathbf{u}^{j})^{T}\,]\,d\upsilon .$$

In (5-34) the indexes n and m of the components of the undisturbed and auxiliary solutions have been omitted. Let us define the solid spherical harmonics as

$$(5\text{-}35)\qquad w_{nm}(r, \vartheta, \lambda) = r^{n}\,Y_{nm}(\vartheta, \lambda).$$

The solid spherical harmonics have the properties:

$$(5\text{-}36)\qquad \nabla^{2}w = \frac{\partial}{\partial x_i}\frac{\partial}{\partial x_i}w = 0$$

$$(5\text{-}37)\qquad \mathbf{r}\cdot\nabla w = x_i\frac{\partial}{\partial x_i}w = n w$$

$$(5\text{-}38)\qquad \mathbf{rr} : \nabla\nabla w = x_i x_k\frac{\partial^{2}}{\partial x_i \partial x_k}w = x_i\frac{\partial}{\partial x_i}\left\{x_k\frac{\partial}{\partial x_k}w\right\} - x_i\,\delta_{ik}\frac{\partial}{\partial x_k}w$$

$$= x_i\frac{\partial}{\partial x_i}\left\{n w\right\} - x_i\frac{\partial}{\partial x_i}w = n(n-1)\,w$$

where $x_i$ (i = 1, 2, 3) indicate Cartesian coordinates, and Einstein's summation convention has been used.

The undisturbed displacement vector $\mathbf{u}^{o}$ and the auxiliary displacement vector $\mathbf{u}^{j}$ can also be expanded in terms of solid spherical harmonics:

(5-39) $$\mathbf{u}^o = h_{n_o} \mathbf{r} w_{n_o m_o} + t_{n_o} \nabla w_{n_o m_o} + \omega_{n_o} \mathbf{r} \times \nabla w_{n_o m_o}$$

(5-40) $$\mathbf{u}^j = h_n^j \mathbf{r} w_{nm}^* + t_n^j \nabla w_{nm}^* + \omega_n^j \mathbf{r} \times \nabla w_{nm}^*$$

where

(5-41) $$h_{n_o} = r^{-(n_o+1)} \left\{ H_{n_o} - n_o T_{n_o} \right\}, \quad t_{n_o} = r^{-(n_o-1)} T_{n_o}, \quad \omega_{n_o} = r^{-n_o} W_{n_o}$$

(5-42) $$h_n^j = r^{-(n+1)} \left\{ H_n^j - n T_n^j \right\}, \quad t_n^j = r^{-(n-1)} T_n^j, \quad \omega_n^j = r^{-n} W_n^j$$

The Cartesian components of the tensor $[ \nabla \mathbf{u}^o + ( \nabla \mathbf{u}^o )^T ]$ can be expressed as

(5-43) $$[ \nabla \mathbf{u}^o + ( \nabla \mathbf{u}^o )^T ]_{ik} = \frac{\partial}{\partial x_i} u_j^o + \frac{\partial}{\partial x_j} u_i^o$$

$$= 2 \dot{h}_{n_o} \frac{x_i}{r} x_k w_{n_o m_o} + 2 h_{n_o} \delta_{ik} w_{n_o m_o} + 2 t_{n_o} \frac{\partial^2}{\partial x_i \partial x_k} w_{n_o m_o} +$$

$$+ ( h_{n_o} + \frac{1}{r} \dot{t}_{n_o} )( x_i \frac{\partial}{\partial x_k} w_{n_o m_o} + x_k \frac{\partial}{\partial x_i} w_{n_o m_o} ) +$$

$$+ \frac{1}{r} \dot{\omega}_{n_o} ( x_i \, \varepsilon_{kpq} x_p \frac{\partial}{\partial x_q} w_{n_o m_o} + x_k \, \varepsilon_{ipq} x_p \frac{\partial}{\partial x_q} w_{n_o m_o} ) +$$

$$+ \omega_{n_o} \left[ \frac{\partial}{\partial x_k} ( \varepsilon_{ipq} x_p \frac{\partial}{\partial x_q} w_{n_o m_o} ) + \frac{\partial}{\partial x_i} \varepsilon_{kpq} x_p \frac{\partial}{\partial x_q} w_{n_o m_o} ) \right]$$

where

(5-44) $$\varepsilon_{kpq} = \mathbf{e}_k \cdot ( \mathbf{e}_p \times \mathbf{e}_q )$$

is called the total asymmetric tensor, or Levi-Civita tensor. The same expression holds for the tensor $[ \nabla \mathbf{u}^j + ( \nabla \mathbf{u}^j )^T ]$, if the indices are changed correspondingly. Using properties (5-36) and (5-37), the tensor inner product $[ \nabla \mathbf{u}^o + ( \nabla \mathbf{u}^o )^T ]$: $[ \nabla \mathbf{u}^j + ( \nabla \mathbf{u}^j )^T ]$ is deduced to be

(5-45) $$[\nabla \mathbf{u}^{o} + (\nabla \mathbf{u}^{o})^{T}] : [\nabla \mathbf{u}^{j} + (\nabla \mathbf{u}^{j})^{T}] =$$

$$= \underline{w_{n_o m_o} w^{*}_{nm}} \cdot 4 \Big\{ [r\dot{h}_{n_o} + (n_o + 1) h_{n_o} + \frac{n_o}{r} \dot{t}_{n_o}] \times$$

$$\times [r\dot{h}^{j}_{n} + (n+1) h^{j}_{n} + \frac{n}{r} \dot{t}^{j}_{n}] + 2 h_{n_o} h^{j}_{n} +$$

$$+ \frac{n(n+1)}{r} \dot{h}_{n_o} t^{j}_{n} + \frac{n_o(n_o+1)}{r} t_{n_o} \dot{h}^{j}_{n} - \frac{n_o n}{2r^2} (r h_{n_o} + \dot{t}_{n_o})(r h^{j}_{n} + \dot{t}^{j}_{n}) \Big\} +$$

$$+ \underline{(\nabla w_{n_o m_o} \cdot \nabla w^{*}_{nm})} \cdot 2r^2 \Big\{ [h_{n_o} + \frac{1}{r} \dot{t}_{n_o} + 2(n_o - 1) \frac{1}{r^2} t_{n_o}] \times$$

$$\times [h^{j}_{n} + \frac{1}{r} \dot{t}^{j}_{n} + 2(n-1) \frac{1}{r^2} t^{j}_{n}] - 4(n_o - 1)(n-1) \frac{1}{r^4} t_{n_o} t^{j}_{n} \Big\} +$$

$$+ \underline{(\nabla\nabla w_{n_o m_o} : \nabla\nabla w^{*}_{nm})} \cdot 4 t_{n_o} t^{j}_{n} +$$

$$+ \underline{(\mathbf{r} \times \nabla w_{n_o m_o}) \cdot (\mathbf{r} \times \nabla w^{*}_{nm})} \cdot \frac{2}{r^2} \Big\{ [r\dot{\omega}_{n_o} + (n_o - 1) \omega_{n_o}] \times$$

$$\times [r\dot{\omega}^{j}_{n} + (n-1) \omega^{j}_{n}] - (n_o - 1)(n-1) \omega_{n_o} \omega^{j}_{n} \Big\} +$$

$$+ \underline{\Big\{ \nabla(\mathbf{r} \times \nabla w_{n_o m_o}) + [\nabla(\mathbf{r} \times \nabla w_{n_o m_o})]^{T} \Big\} : \Big\{ \nabla(\mathbf{r} \times \nabla w^{*}_{nm}) +}$$

$$\underline{+ [\nabla(\mathbf{r} \times \nabla w^{*}_{nm})]^{T} \Big\}} \cdot \omega_{n_o} \omega^{j}_{n} + \underline{[\mathbf{r} \cdot (\nabla w_{n_o m_o} \times \nabla w^{*}_{nm})]} \times$$

$$\times 2 \Big\{ (h^{j}_{n} + \frac{1}{r} \dot{t}^{j}_{n}) [r\dot{\omega}_{n_o} + (n_o - 1) \omega_{n_o}] + 2(n-1) \frac{1}{r} \dot{\omega}_{n_o} t^{j}_{n} -$$

$$- (h_{n_o} + \frac{1}{r} \dot{t}_{n_o}) [r\dot{\omega}^{j}_{n} + (n-1) \omega^{j}_{n}] - 2(n_o - 1) \frac{1}{r} \dot{\omega}^{j}_{n} t_{n_o} \Big\} +$$

$$+ \underline{[\nabla\nabla w_{n_o m_o} : \nabla(\mathbf{r} \times \nabla w^{*}_{nm})]} \cdot 4 \omega_{n_o} t^{j}_{n} +$$

$$+ \underline{[\nabla\nabla w^{*}_{nm} : \nabla(\mathbf{r} \times \nabla w_{n_o m_o})]} \cdot 4 \omega^{j}_{n} t_{n_o}$$

Here we use the cross symbol $\times$ for both the vector product and the normal multiplication. Using the results for the integrals with integrands denoted by ~~~, given in Appendix B, and transforming inversely variables h, t and $\omega$ to variables H, T and W, respectively, we obtain the semi-analytical solution of integral (5-34) as

$$\Gamma^{j}_{nm}(\delta\mu) = -\iiint_{r\le a} \delta\mu [\nabla\mathbf{u}^{o} + (\nabla\mathbf{u}^{o})^{T}] : \nabla\mathbf{u}^{j}\, dv \tag{5-35}$$

$$= \sum_{l}\sum_{k} \Big\{ \tilde{A}(n_o, m_o; n, m; l, k) \int_0^a \mu_{lk} K^{j}_{\mu}(n_o, n, l, r)\, dr$$

$$+ i\tilde{B}(n_o, m_o; n, m; l, k) \int_0^a \mu_{lk} G^{j}_{\mu}(n_o, n, l, r)\, dr \Big\}$$

where

$$K^{j}_{\mu} = -2r^2 \dot{H}_{n_o}\dot{H}^{j}_{n} - 2H_{n_o}(H^{j}_{n} - n(n+1)T^{j}_{n}) -$$

$$-2H^{j}_{n}(H_{n_o} - n_o(n_o+1)T_{n_o}) + n_o n(H_{n_o} + r\dot{T}_{n_o} - T_{n_o})\times$$

$$\times(H^{j}_{n} + r\dot{T}^{j}_{n} - T^{j}_{n}) - \frac{1}{2}(n_o + n + l + 1)(n_o + n - l)\times$$

$$\times[(H_{n_o} + r\dot{T}_{n_o} - T_{n_o})(H^{j}_{n} + r\dot{T}^{j}_{n} - T^{j}_{n}) - 4(n_o - 1)(n - 1)T_{n_o}T^{j}_{n}]$$

$$- \frac{1}{2}[(n_o + n + l)^2 - 1](n_o + n - l)(n_o + n - l - 2)T_{n_o}T^{j}_{n}$$

$$-[\frac{1}{2}(n_o + n + l + 1)(n_o + n - l) - n_o n][(r\dot{W}_{n_o} - W_{n_o})\times$$

$$\times(r\dot{W}^{j}_{n} - W^{j}_{n}) - (n_o - 1)(n - 1)W_{n_o}W^{j}_{n}] - \frac{1}{2}(n_o + n + l + 1)\times$$

$$\times(n_o + n - l)[(n_o + n + l + 1)(n_o + n - l) -$$

$$-(3n_o n + n_o + n + 1)]W_{n_o}W^{j}_{n}$$

$$G^{j}_{\mu} = (r\dot{W}_{n_o} - W_{n_o})(H^{j}_{n} + r\dot{T}^{j}_{n} - T^{j}_{n}) -$$

$$-(H_{n_o} + r\dot{T}_{n_o} - T_{n_o})(r\dot{W}^{j}_{n} - W^{j}_{n}) + [2(n_o - 1)(n - 1) +$$

$$+(n_o + n + l)(n_o + n - l - 1)](T_{n_o}W^{j}_{n} - W_{n_o}T^{j}_{n})$$

and $\tilde{A}$ and $\tilde{B}$ with indices $(n_o, m_o; n, m; l, k)$ are expansion coefficients given in Appendix B. This formula can be used directly for numerical programming.

## Chapter 6. Geophysical Observables at the Disturbed Earth's Surface

In the previous chapter we have determined the variations of tidal displacement due to different earth's parameter perturbations:

$$\delta \mathbf{u} = \sum_n \sum_m \left\{ H_n^m \, \mathbf{e}_r + T_n^m \, \nabla_1 + W_n^m \, \mathbf{e}_r \wedge \nabla_1 \right\} Y_{nm} (\vartheta, \lambda) \, e^{i\omega t}. \tag{6-1}$$

Strictly speaking, these variations are related to the inside of the undisturbed earth's surface. We shall use $\Delta$ to represent the difference between any disturbed quantity evaluated at the disturbed boundary and the corresponding undisturbed quantity evaluated at the undisturbed boundary. The variation $\Delta f$ to any quantity f is thus given by

$$\Delta f \Big|^{+,-} = \delta f \Big|^{+,-} + \delta h \frac{\partial f^o}{\partial r} \Big|^{+,-} \tag{6-2}$$

where $\delta h$ is the boundary undulation. If f is a continuous quantity through the boundary, we have

$$\Delta f = \Delta f \Big|^{+} = \Delta f \Big|^{-}. \tag{6-3}$$

In the following, the variation represented by $\Delta$ will be called the total variation, in order to distinguish it from the variation represented by $\delta$ at points fixed in the space. The total variation of the normal of the earth's surface is given by

$$\Delta \mathbf{n} = -\frac{1}{a} \nabla_1 \delta h. \tag{6-4}$$

As a good approximation, we take in the following the geoid as the earth's surface so that the geometrical normal $\mathbf{n}$ coincides with the gravity normal $-\nabla V / g$ at the undeformed earth's surface.

### 6.1 Total variation in displacement

Omitting the time factor $\exp(i\omega t)$, the total variation of the displacement field is given by

$$\Delta \mathbf{u} = \delta \mathbf{u} \Big|^{-} + \delta h \frac{\partial \mathbf{u}^o}{\partial r} \Big|^{-}. \tag{6-5}$$

(a) Vertical component $\Delta u^{(n)}$ (upward positive)

It is convenient to give the components of all geophysical observables in the local Cartesian coordinates. The vertical component of the total variation of the displacement $\Delta u^{(n)}$ can be expressed as

$$(6\text{-}6) \qquad \Delta u^{(n)} = \Delta(\mathbf{n}\cdot\mathbf{u}) = \Delta\mathbf{n}\cdot\mathbf{u}^{o} + \mathbf{n}^{o}\cdot\Delta\mathbf{u}$$

$$= -\frac{1}{a}\nabla_1 \delta h\cdot\mathbf{u}^{o} + \mathbf{e}_r\cdot\Delta\mathbf{u}$$

$$= \sum_n\sum_m \Delta U_{nm} Y_{nm}$$

where

$$\Delta U_{nm} = H_n^m + \frac{1}{a}\sum_l\sum_k h_{lk}\left\{(a\dot{H}_{n_o} - p_{n_o l} T_{n_o})\tilde{A} - i W_{n_o}\tilde{B}\right\}$$

$$p_{n_o l} = \frac{1}{2}(n_o + n + l + 1)(n_o + l - n) - n_o l$$

$$i = \sqrt{-1}\ .$$

The expansion coefficients $\tilde{B}$ and $\tilde{A}$ with the arguments ( $n_o$, $m_o$; n, m; l, k ) are given in Appendix B.

(b) North-south component $\Delta u^{(s)}$ (southward positive)

It can be shown that the total variation of the southward unit tangential vector of the earth's surface is

$$(6\text{-}7) \qquad \Delta\mathbf{e}_s = \frac{1}{a}\mathbf{e}_r\frac{\partial}{\partial\vartheta}\delta h.$$

The north-south component of the total variation of the displacement $\Delta u^{(s)}$ is thus given by

$$(6\text{-}8) \qquad \Delta u^{(s)} = \Delta(\mathbf{e}_s\cdot\mathbf{u}) = \Delta\mathbf{e}_s\cdot\mathbf{u}^{o} + \mathbf{e}_s^{o}\cdot\Delta\mathbf{u} = \frac{1}{a}u_r^{o}\frac{\partial}{\partial\vartheta}\delta h + \mathbf{e}_\vartheta\cdot\Delta\mathbf{u}$$

$$= \sum_n\sum_m\left\{\Delta V_{nm}\frac{\partial}{\partial\vartheta} - \Delta W_{nm}\frac{1}{\sin\vartheta}\frac{\partial}{\partial\lambda}\right\}Y_{nm}$$

with

$$\Delta V_{nm} = T_n^m + \frac{1}{n(n+1)a} \sum_l \sum_k h_{lk} \left\{ (a p_{n_o n} \dot{T}_{n_o} + p_{nl} H_{n_o}) A + i \dot{W}_{n_o} \tilde{B} \right\}$$

$$\Delta W_{nm} = W_n^m + \frac{1}{n(n+1)a} \sum_l \sum_k h_{lk} \left\{ a p_{n_o n} \dot{W}_{n_o} A - i ( a \dot{T}_{n_o} + H_{n_o}) \tilde{B} \right\}$$

$$p_{n_o n} = \frac{1}{2} (n_o + n + l + 1)(n_o + n - l) - n_o n$$

$$p_{nl} = \frac{1}{2} (n_o + n + l + 1)(n + l - n_o) - nl.$$

(c) East-west component $\Delta u^{(e)}$ (eastward positive)

The total variation of the eastward unit tangential vector of the earth's surface is given by

(6-9) $$\Delta \mathbf{e}_e = \frac{\mathbf{e}_r}{a \sin\vartheta} \frac{\partial}{\partial \lambda} \delta h,$$

and the east-west component of the total variation of the displacement is expressed by

(6-10) $$\Delta u^{(e)} = \Delta(\mathbf{e}_e \cdot \mathbf{u}) = \Delta \mathbf{e}_e \cdot \mathbf{u}^o + \mathbf{e}_e^o \cdot \Delta \mathbf{u} = \frac{u_r^o}{a \sin\vartheta} \frac{\partial}{\partial \lambda} \delta h + \mathbf{e}_\vartheta \cdot \Delta \mathbf{u}$$

$$= \sum_n \sum_m \left\{ \Delta V_{nm} \frac{1}{\sin\vartheta} \frac{\partial}{\partial \lambda} + \Delta W_{nm} \frac{\partial}{\partial \vartheta} \right\} Y_{nm}$$

where $\Delta V_{nm}$ and $\Delta W_{nm}$ are given by (6-8).

6.2 Variation in free space Eulerian potential

The potential variation in the free space has been determined in the previous chapter and is given by

(6-11) $$\delta\varphi = \sum_n \sum_m \delta R_{nm} \left( \frac{a}{r} \right)^{n+1} Y_{nm}(\vartheta, \lambda), \quad \text{for } r > a$$

where

$$\delta R_{nm} = R_n^m + 4\pi G\rho^o H_{n_o} \sum_l \sum_k \tilde{A}(n_o, m_o; n, m; l, k)\, h_{lk}.$$

Note that in free space only the variation of the Eulerian potential at fixed points is of interest. It would disturb the satellite orbits and could be observed by, for example, Satellite Laser Ranging (SLR).

### 6.3 Total variation in tidal gravity (downward positive)

The tidal acceleration $\mathbf{g}_t$ at the deformed earth's surface is given by

$$\mathbf{g}_t = \nabla\varphi\big|^+ + \mathbf{u}\cdot\nabla\nabla V\big|^+ - \ddot{\mathbf{u}} - 2\mathbf{\Omega}\times\dot{\mathbf{u}}. \tag{6-12}$$

The latter two terms are the tidally induced inertial force and the Coriolis force. The centrifugal force has been included in the second term by introducing the perturbation of the earth's gravity field. Similar to the treatment used in the previous section, the total variation of the tidal acceleration can be to first order expressed as

$$\Delta\mathbf{g}_t = \nabla\delta\varphi\big|^+ + \delta h\frac{\partial}{\partial r}\nabla\varphi^o\big|^+ + \mathbf{u}^o\cdot\delta h\frac{\partial}{\partial r}\nabla\nabla V^o\big|^+ + \mathbf{u}^o\cdot\nabla\nabla\delta\Phi\big|^+ + $$

$$+ \Delta\mathbf{u}\cdot\nabla\nabla V^o\big|^+ + \omega^2\mathbf{u}^o - 2i\omega\mathbf{\Omega}\times\mathbf{u}^o - \mathbf{\Omega}\times(\mathbf{\Omega}\times\mathbf{u}^o).$$

It can be shown that,

$$\nabla\delta\varphi\big|^+ = \sum_n\sum_m \delta R_{nm}\frac{1}{a}[-(n+1)Y_{nm}\mathbf{e}_r + \nabla_1 Y_{nm}] \tag{6-13}$$

$$\frac{\partial}{\partial r}\nabla\varphi^o\big|^+ = \frac{R_{n_o}}{a^2(1+k_{n_o})}\Big\{[n_o(n_o-1) + (n_o+1)(n_o+2)k_{n_o}]\mathbf{e}_r + \tag{6-14}$$

$$+ [(n_o-1) - (n_o+2)k_{n_o}]\nabla_1\Big\}Y_{n_o m_o}$$

$$\nabla\nabla V^o\big|^+ = \frac{g^o}{a}(3\mathbf{e}_r\mathbf{e}_r - \mathbf{I}) \tag{6-15}$$

$$\Delta\mathbf{u}\cdot\nabla\nabla V^o\big|^+ = \frac{g^o}{a}(3\Delta u_r\mathbf{e}_r - \Delta\mathbf{u}) \tag{6-16}$$

$$(6\text{-}17)\qquad \frac{\partial}{\partial r}\nabla\nabla \mathbf{V}^{o}\Big|^{+} = -\frac{3}{a}\nabla\nabla \mathbf{V}^{o}\Big|^{+}$$

where all quantities are related to the undisturbed earth's surface, and

$$(6\text{-}18)\qquad \delta\Phi\Big|^{+} = \sum_{l}\sum_{k} \Phi_{lk}\left(\frac{a}{r}\right)^{l+1} Y_{lk}.$$

Usually, the vertical component of the tidal acceleration as measured by a gravimeter, is called the tidal gravity. Its total variation can be expressed as

$$\Delta g_t^{(n)} = -\mathbf{n}^{o}\cdot\Delta\mathbf{g}_t - \Delta\mathbf{n}\cdot\mathbf{g}_t^{o} = -\mathbf{e}_r\cdot\Delta\mathbf{g}_t + \frac{1}{a}\nabla_1\delta h\cdot\mathbf{g}_t^{o}$$

where

$$\mathbf{g}_t^{o} = \nabla\varphi^{o}\Big|^{+} + \mathbf{u}^{o}\cdot\nabla\nabla V^{o}\Big|^{+}$$

is the undisturbed tidal gravity at the spherical earth's surface.

Using a similar technique as above, we obtain

$$(6\text{-}19)\qquad \Delta g_t^{(n)} = \sum_{n}\sum_{m} \Delta g_{nm} Y_{nm}$$

with

$$\Delta g_{nm} = \frac{(n+1)}{a} R_n^m - \frac{2g^o}{a} H_n^m - \left(\frac{2}{3}\Omega^2 + \omega^2\right) H_{n_o} \delta_{nn_o} \delta_{mm_o} +$$

$$+ \frac{1}{a^2} \sum_l \sum_k \Big\{ h_{lk} \Big[ \Big( p_{n_o l} (R_{n_o} - g^o T_{n_o}) + g^o (6H_{n_o} - 2a\dot{H}_{n_o}) +$$

$$- \frac{n_o(n_o-1) + (n_o+1)(n_o+2)k_{n_o}}{(1+k_{n_o})} R_{n_o} + 4\pi G \rho^o (n_o+1) H_{n_o} \Big) \tilde{A} -$$

$$- i g^o W_{n_o} \tilde{B} \Big] + \frac{\Phi_{lk}}{a^2} \Big[ \big( -(l+1)(l+3) H_{n_o} + (l+2) p_{n_o l} T_{n_o} \big) \tilde{A} +$$

$$+ i(l+2) W_{n_o} \tilde{B} \Big] + \sqrt{\frac{16\pi}{3}}\, \Omega \omega \delta_{l1} \delta_{ko} \Big[ T_{n_o} \tilde{B} + i p_{n_o l} W_{n_o} \tilde{A} \Big] +$$

$$+ \sqrt{\frac{4\pi}{45}}\, \Omega^2 \delta_{l2} \delta_{ko} \Big[ (l H_{n_o} + p_{n_o l} T_{n_o}) \tilde{A} + i W_{n_o} \tilde{B} \Big] \Big\}.$$

## 6.4 Total variation in tilt

The tidal tilt vector $\mathbf{t}_t$, measured by a tiltmeter at the deformed earth's surface, is defined as the horizontal projection of the spherical angle between the instantaneous geometrical normal and the instantaneous outword gravity normal:

$$\text{(6-20)} \qquad \mathbf{t}_t = (\mathbf{I} - \mathbf{nn}) \cdot \Big\{ \frac{\mathbf{g}_t}{g} + \mathbf{u} \cdot (\mathbf{I} - \mathbf{nn}) \cdot \nabla \mathbf{n} - \nabla (\mathbf{n} \cdot \mathbf{u}) \Big\}$$

where the first term is the horizontal projection of the outward tidal gravity normal, while the last two terms describe the negative variation of the geometrical surface normal due to the horizontal and the vertical displacement, respectively.

(a) North-south component $\Delta t_t^{(s)}$ (southward positive)

The total variation of the north-south component of tilt is determined by

$$\Delta t_t^{(s)} = \Delta \{ \mathbf{e}_s \cdot \mathbf{t}_t \}$$

Using a similar technique as above, it can be expressed in the form

$$\text{(6-21)} \qquad \Delta t_t^{(s)} = \sum_n \sum_m \Big\{ \Delta t_{nm} \frac{\partial}{\partial \vartheta} - \Delta \omega_{nm} \frac{1}{\sin \vartheta} \frac{\partial}{\partial \lambda} \Big\} Y_{nm}$$

where

$$\Delta t_{nm} = \frac{1}{ag^{o}} R_n^m - \frac{1}{a} H_n^m + \frac{\omega^2}{g^{o}} T_{n_o} \delta_{nn_o} \delta_{mm_o} +$$

$$+ \sum_{l}\sum_{k} \Big\{ \frac{1}{a} h_{lk} \Big[ \Big( \frac{4\pi G \rho^{o}}{g^{o}} H_{n_o} - \dot{H}_{n_o} + p_{n_o l} \frac{1}{a} T_{n_o} +$$

$$+ \frac{p_{n_o n}}{n(n+1)} \Big( \frac{1}{g^{o}} \dot{R}^{+}_{n_o} - \frac{1}{ag^{o}} R_{n_o} + \frac{1}{a} H_{n_o} \Big) +$$

$$+ \frac{p_{nl}}{n(n+1)} \Big( \frac{1}{g^{o}} \dot{R}_{n_o} - \dot{H}_{n_o} \Big) \Big) \tilde{A} + i \frac{1}{a} W_{n_o} \tilde{B} \Big] -$$

$$- \frac{1}{n(n+1)} \Big[ \Big( \frac{1}{g^{o}} \Delta G_{lk} \Big( \frac{p_{n_o n}}{ag^{o}} R_{n_o} + \frac{p_{nl}}{a} H_{n_o} \Big) +$$

$$+ \frac{p_{nl}}{a^2 g^{o}} \Phi_{lk} H_{n_o} - \sqrt{\frac{4\pi}{45}} \frac{\Omega^2 p_{nl}}{g^{o}} H_{n_o} \delta_{12} \delta_{k0} \Big) \tilde{A} -$$

$$- \sqrt{\frac{16\pi}{3}} \frac{\Omega\omega}{g^{o}} \Big( (H_{n_o} - T_{n_o}) \tilde{B} + i W_{n_o} \tilde{A} \Big) \delta_{11} \delta_{k0} \Big] \Big\}$$

$$\Delta\omega_{nm} = \frac{\omega^2}{g^{o}} W_{n_o} \delta_{nn_o} \delta_{mm_o} +$$

$$+ \frac{i}{n(n+1) g^{o}} \sum_{l}\sum_{k} \Big\{ \Big[ - \frac{1}{a^2} h_{lk} R_{n_o} - \Delta G_{lk} \Big( \frac{1}{ag^{o}} R_{n_o} - \frac{1}{a} H_{n_o} \Big)$$

$$+ \frac{1}{a^2} \Phi_{lk} H_{n_o} - \sqrt{\frac{4\pi}{45}} \Omega^2 H_{n_o} \delta_{12} \delta_{k0} \Big] \tilde{B} +$$

$$+ \sqrt{\frac{16\pi}{3}} \Omega\omega \Big[ \frac{1}{a} ( p_{nl} H_{n_o} - p_{n_o n} T_{n_o} ) \tilde{A} + \frac{1}{a} W_{n_o} i \tilde{B} \Big] \delta_{11} \delta_{k0} \Big\}$$

and

$$\Delta G_{lk} = (l+1) \Phi_{lk} / a - 2 g^{o} h_{lk} / a + \sqrt{\frac{16\pi}{9}} \Omega^2 a ( \delta_{12} \delta_{k0} / \sqrt{5} - \delta_{10} \delta_{k0} )$$

$$\dot{R}^{+}_{n_o} = \dot{R}_{n_o} - 4\pi G \rho^{o} H_{n_o} .$$

(b) East-west component $\Delta t_t^{(e)}$ (eastward positive)

$$\Delta t_t^{(e)} = \sum_n \sum_m \left\{ \Delta t_{nm} \frac{1}{\sin\vartheta} \frac{\partial}{\partial\lambda} + \Delta\omega_{nm} \frac{\partial}{\partial\vartheta} \right\} Y_{nm} \tag{6-22}$$

where $\Delta t_{nm}$ and $\Delta\omega_{nm}$ are given above.

6.5 Station dependent tidal parameters

It is convenient to express the tidal responses of the earth by some dimensionless coefficients. These coefficients are usually defined as the ratio of the tidal responses in displacement, gravity, tilt and so on, measured at the earth's surface to the corresponding quantities calculated for a rigid earth.

The gravimetric factor $\delta$, for example, is defined as the ratio of the tidally induced gravity as measured by a gravimeter at the deformed earth's surface to the gravitational acceleration caused directly by the tide-generating bodies. The latter is usually called the theoretical tidal gravity. For a spherically symmetric, non-rotating earth, the gravimeter factor depends only upon the degree of spherical harmonics. However, if we consider spherically asymmetric perturbations e.g. rotation, lateral heterogeneities, and interior boundary undulations, and the surface (geoid) undulations, the gravimetric factor depends not only on the degree, but also on the order of spherical harmonics and, in general, also on the location of stations. Wahr (1979) has determined the gravimetric factor for a rotating, elliptical and elastic earth, which depends on the tidal type (degree and order of the harmonic expansion) and the latitude but does not depend on the longitude because of the axial symmetry. In his paper, the theoretical tidal gravity is related to the undisturbed spherical earth's surface. Dehant (1987) pointed out that if the gravimetric factor is to be understood as a transfer function, the theoretical tidal gravity should be related to the disturbed elliptical earth's surface. Although the gravimetric factor redefined by Dehant has the correct physical meaning as a transfer function, its calculation becomes much more complicated and unpractical for a more realistic earth surface such as the geoid. Here, we consider the tidal parameters to be not more than some normalized geophysical quantities. It is obvious, that the simpler the reference surface for the corresponding theoretical tidal quantities is, the more practical is the normalization. Furthermore, one should take into account the traditional convention in the definition of the tidal parameters. For this reason, here we prefer Wahr's definition.

Without loss of generality, we assume the tide-generating potential taking the complex form

$$\varphi^P(n_o, m_o; \omega, \mathbf{r}) = C_{n_o}^{m_o}(\omega) \left(\frac{r}{a}\right)^{n_o} Y_{n_o m_o}(\vartheta, \lambda)\, e^{i\omega t} \tag{6-23}$$

where coefficient $C^{m_o}_{n_o}$ determins the amplitude and phase of the potential. For simplicity, we will omit the time dependent factor $\exp(i\omega t)$.

(a) Love number h

As stated above, the tidal parameters are in general dependent on the tidal type, the tidal frequency and the location of stations. The equivalent Love number h is defined as the ratio of the tidally induced vertical displacement at the real earth's surface to the equilibrium tide which describes the tidally induced displacement of the equipotential surface of the spherically symmetric rigid earth:

$$
\begin{aligned}
(6\text{-}24) \qquad h(n_o, m_o; \omega, \vartheta, \lambda) &= \frac{u^{(n)}}{\varphi^P / g^o \big|_{r=a}} \\
&= \frac{u^o_r + \Delta u^{(n)}}{\varphi^P / g^o \big|_{r=a}} \\
&= h_{n_o} + \Delta h(n_o, m_o; \omega, \vartheta, \lambda)
\end{aligned}
$$

where

$$(6\text{-}25) \qquad h_{n_o} = H_{n_o} / \zeta^o \, , \quad \text{with} \quad \zeta^o = C^{m_o}_{n_o} / g^o$$

is the undisturbed Love number h determined for the undisturbed earth SNREI, and

$$(6\text{-}26) \qquad \Delta h(n_o, m_o; \omega, \vartheta, \lambda) = \sum_n \sum_m \Delta U_{nm}(\omega) Y_{nm}(\vartheta, \lambda) \Big/ \left( \zeta^o Y_{n_o m_o}(\vartheta, \lambda) \right)$$

is the total variation of the Love number h.

(b) Love number l

In the general case, the Love number l will be anisotropic. In the north-south direction the equivalent Love number $l^{(s)}$ is defined by

$$(6\text{-}27) \qquad l^{(s)}(n_o, m_o; \omega, \vartheta, \lambda) = u^{(s)} \Big/ \left( \zeta^o \frac{\partial}{\partial \vartheta} Y_{n_o m_o}(\vartheta, \lambda) \right)$$

and in the west-east direction

(6-28) $$l^{(e)}(n_o, m_o; \omega, \vartheta, \lambda) = u^{(e)} \Big/ \Big( \frac{\zeta^o}{\sin\vartheta} \frac{\partial}{\partial\lambda} Y_{n_o m_o}(\vartheta, \lambda) \Big).$$

Using the expressions

(6-29) $$l^{(s,e)}(n_o, m_o; \omega, \vartheta, \lambda) = l_{n_o} + \Delta l^{(s,e)}(n_o, m_o; \omega, \vartheta, \lambda)$$

where

(6-30) $$l_{n_o} = T_{n_o} / \zeta^o$$

is the undisturbed Love number l, we obtain from (6-8) and (6-10)

(6-31) $$\Delta\, l^{(s)}(n_o, m_o; \omega, \vartheta, \lambda)$$

$$= \sum_n \sum_m \Big\{ \Delta V_{nm} \frac{\partial}{\partial\vartheta} - \Delta W_{nm} \frac{1}{\sin\vartheta} \frac{\partial}{\partial\lambda} \Big\} Y_{nm} \Big/ \Big( \zeta^o \frac{\partial}{\partial\vartheta} Y_{n_o m_o} \Big)$$

and

(6-32) $$\Delta\, l^{(e)}(n_o, m_o; \omega, \vartheta, \lambda)$$

$$= \sum_n \sum_m \Big\{ \Delta V_{nm} \frac{1}{\sin\vartheta} \frac{\partial}{\partial\lambda} + \Delta W_{nm} \frac{\partial}{\partial\vartheta} \Big\} Y_{nm} \Big/ \Big( \zeta^o \frac{i m_o}{\sin\vartheta} Y_{n_o m_o} \Big).$$

(c) Love number k

The total variation of the Eulerian potential can be formulated in a way similar to the one used by Dehant (1987) for the displacement field related to the earth's surface. She redefined the Love number k as the transfer function between the tide-generating potential and the Eulerian potential associated with the tidal deformations, both related to the elliptical earth's surface. However, we think this treatment is not practical because, at present, no terrestrial technique can contribute to directly measuring this parameter. We prefer to define the Love number k as related to the undisturbed spherical surface because this is convenient for extending the variation of the Eulerian potential to free space. The variation of the potential at the fixed points related to the outside of the undisturbed earth's surface has been given in the previous chapter as

$$\delta\varphi\Big|^{+} = \sum_{n}\sum_{m} \delta R_{nm} Y_{nm}(\vartheta, \lambda), \qquad \text{for } r = a^{+}$$

and in free space

$$\delta\varphi = \sum_{n}\sum_{m} \delta R_{nm} \left(\frac{a}{r}\right)^{n+1} Y_{nm}(\vartheta, \lambda), \qquad \text{for } r > a.$$

This variation is entirely associated with the perturbation of tidal deformations in the earth's interior. The equivalent Love number k is defined as the ratio of the deformation part of the potential $\varphi^{s}$ related to the spherical mean earth's surface to the tide-generating potential $\varphi^{P}$ related to the same level:

$$\text{(6-33)} \qquad k(n_o, m_o; \omega, \vartheta, \lambda) = \frac{\varphi^{s}}{\varphi^{P}}\Big|_{r=a^{+}}$$

$$= \frac{\varphi^{os} + \delta\varphi}{\varphi^{P}}\Big|_{r=a^{+}}$$

Using

$$\text{(6-34)} \qquad k(n_o, m_o; \omega, \vartheta, \lambda) = k_{n_o} + \delta k(n_o, m_o; \omega, \vartheta, \lambda)$$

we obtain

$$\text{(6-35)} \qquad k_{n_o} = \frac{\varphi^{os}}{\varphi^{P}}\Big|_{r=a} = \frac{R_{n_o}}{g^{o}\zeta^{o}} - 1$$

as the undisturbed Love number k, and

$$\text{(6-36)} \qquad \delta k(n_o, m_o; \omega, \vartheta, \lambda) = \frac{\delta\varphi}{\varphi^{P}}\Big|_{r=a^{+}} = \sum_{n}\sum_{m} \delta R_{nm} Y_{nm} \Big/ \left(g^{o}\zeta^{o} Y_{n_o m_o}\right)$$

as the variation of the Love number k, related to the spherical mean earth's surface.

(d) Gravimetric factor $\delta$

The gravimetric factor is defined as the ratio of the tidally induced gravity, measured by a gravimeter at the deformed real earth's surface, to the theoretical tidal force, evaluated at the corresponding point on the spherical mean earth's surface:

$$\text{(6-37)} \quad \delta(n_o, m_o; \omega, \vartheta, \lambda) = g_t^{(n)} \Big/ \Big[ -\frac{\partial}{\partial r} \varphi^P \Big]_{r=a}$$

$$= \left( \mathbf{e}_r \cdot \mathbf{g}_t^o + \Delta g_t^{(n)} \right) \Big/ \Big[ -\frac{\partial}{\partial r} \varphi^P \Big]_{r=a}$$

$$= \delta_{n_o} + \Delta\delta(n_o, m_o; \omega, \vartheta, \lambda)$$

where

$$\text{(6-38)} \quad \delta_{n_o} = 1 + \frac{2}{n_o} h_{n_o} - \frac{n_o + 1}{n_o} k_{n_o}$$

is the undisturbed gravimetric factor, and

$$\text{(6-39)} \quad \Delta\delta(n_o, m_o; \omega, \vartheta, \lambda) = -\frac{a(1 + k_{n_o})}{n_o R_{n_o}} \sum_n \sum_m \Delta g_{nm} Y_{nm} \Big/ Y_{n_o m_o}$$

is its total variation.

(e) Tilt factor $\gamma$

In the same way as the equivalent Love number l, the equivalent tilt factor $\gamma$ will become anisotropic in the general case. In the north-south direction it is defined as

$$\text{(6-40)} \quad \gamma^{(s)}(n_o, m_o; \omega, \vartheta, \lambda) = a\, t_t^{(s)} \Big/ \Big( \zeta^o \frac{\partial}{\partial \vartheta} Y_{n_o m_o}(\vartheta, \lambda) \Big)$$

and in the west-east direction as

$$\text{(6-41)} \quad \gamma^{(e)}(n_o, m_o; \omega, \vartheta, \lambda) = a\, t_t^{(e)} \Big/ \Big( \frac{\zeta^o}{\sin\vartheta} \frac{\partial}{\partial \lambda} Y_{n_o m_o}(\vartheta, \lambda) \Big).$$

Using the expressions

$$\gamma^{(s,e)}(n_o, m_o; \omega, \vartheta, \lambda) = \gamma_{n_o} + \Delta\gamma^{(s,e)}(n_o, m_o; \omega, \vartheta, \lambda) \tag{6-42}$$

and with a procedure similar to the one used in section (b), we obtain

$$\gamma_{n_o} = 1 + k_{n_o} - h_{n_o} \tag{6-43}$$

as the undisturbed tilt factor, and, using (6-21) and (6-22),

$$\Delta\gamma^{(s)}(n_o, m_o; \omega, \vartheta, \lambda)$$
$$= \sum_n \sum_m \left\{ \Delta t_{nm} \frac{\partial}{\partial\vartheta} - \Delta\omega_{nm} \frac{1}{\sin\vartheta} \frac{\partial}{\partial\lambda} \right\} Y_{nm} \Big/ \left( \frac{\zeta^o}{a} \frac{\partial}{\partial\vartheta} Y_{n_o m_o} \right) \tag{6-44}$$

and

$$\Delta\gamma^{(e)}(n_o, m_o; \omega, \vartheta, \lambda)$$
$$= \sum_n \sum_m \left\{ \Delta t_{nm} \frac{1}{\sin\vartheta} \frac{\partial}{\partial\lambda} + \Delta\omega_{nm} \frac{\partial}{\partial\vartheta} \right\} Y_{nm} \Big/ \left( \frac{\zeta^o}{a} \frac{i m_o}{\sin\vartheta} Y_{n_o m_o} \right) \tag{6-45}$$

as the total variations of the tilt factor.

It can be seen that the modification of the irregular earth's parameter perturbations on earth tides can be characterized by the following main effects in the equivalent tidal parameters:

— *Dependence on earth tide type*: The equivalent tidal parameters depend not only on the degree but also on the order of the harmonic term of the tide-generating potential, i.e. they are different for the sectorial, tesseral and zonal tides;

— *Dependence on the station coordinates*: Because of the asymmetric properties of the earth's parameter distribution the equivalent tidal parameters are not constant but dependent on both latitude and longitude;

— *Dependence on the tidal frequency*: The equivalent tidal parameters are frequency dependent because of the anelastic dispersion of the Love numbers and the effect of inertial forces;

— *Phase shift*: The equivalent tidal parameters become complex because the tidal response of the earth shows phase shifts to the tide-generating force. This effect is attributable not only to the anelastic dispersion but also to the asymmetric parameter distribution;

— *Anisotropy*: The equivalent tidal parameters, which describe the vectorial observables, for example, horizontal displacement and tilt, become anisotropic. This is also attributable to the asymmetric earth's parameter distribution;

— *Infinity*: Because the nodal lines of the tidal response on the earth's surface are in general different from that of the theoretical tides, the equivalent tidal parameters may become infinite.

# Chapter 7. Numerical Results

## 7.1 Unit system for computational procedure

In order to avoid operating with too large or too small quantities in the numerical procedure, all formulae will be made dimensionless by using the unit system, in which the earth's mean radius ($a$ = 6371 km) is taken as the unit of length, the mean density of the earth ($\bar{\rho}^o$ = 5.517 g/cm$^3$) as the unit of density, and the mean gravitational acceleration at the earth's surface ($g^o(a)$ = 981.96 cm/s$^2$) as the unit of acceleration. In this system all other units can be deduced from these three, for example:

| | |
|---|---|
| **Length** | $\mathbf{a}$ |
| **Density** | $\bar{\rho}^o$ |
| **Acceleration** | $\mathbf{g^o(a)}$ |
| Mass | $a^3\,\bar{\rho}^o$ |
| Time | $[a/g^o(a)]^{1/2}$ |
| Velocity | $[a g^o(a)]^{1/2}$ |
| Frequency | $[g^o(a)/a]^{1/2}$ |
| Potential | $a g^o(a)$ |
| Pressure (elastical modulus) | $a g^o(a)\,\bar{\rho}^o$ |
| Gravitational constant | $3/(4\pi)$ |

## 7.2 Examining the perturbation method for special cases

In order to check the validity of our theoretical formulizations and computational programming, it is necessary to compare the numerical results with the analytical results of some special cases.

(a) Statical tidal deformations on an elliptical, rotating, incompressible, homogeneous and fluid earth

In this simple case, the solution can be determined analytically by using the following three conditions:

—— Conservation of the total mass;
—— Equipotential earth's surface;
—— No density redistribution in the earth's interior because of incompressibility.

The undeformed elliptical earth's surface can be described to first order by the equation

(7-1) $$r_o = 1 - \frac{2}{3}\varepsilon P_2(\cos\vartheta)$$

where $\varepsilon$ is the ellipticity. We assume that the tide-generating potential is expressed by

(7-2) $$\varphi^P(r, \vartheta, \lambda) = \varphi^o r^2 Y_{2m}(\vartheta, \lambda), \qquad m = 0, 1, 2.$$

and suppose that the vertical displacement of the earth's surface takes the form

(7-3) $$u^{(n)}(\vartheta, \lambda) = \varphi^o\left\{\delta h_o Y_{Om}(\vartheta, \lambda) + h_2 Y_{2m}(\vartheta, \lambda) + \delta h_4 Y_{4m}(\vartheta, \lambda)\right\}$$

and the deformation-induced Eulerian potential in the earth's exterior

(7-4) $$\varphi^s(r, \vartheta, \lambda) = \varphi^o\left\{\frac{\delta k_o}{r} Y_{Om}(\vartheta, \lambda) + \frac{k_2}{r^3} Y_{2m}(\vartheta, \lambda) + \frac{\delta k_4}{r^5} Y_{4m}(\vartheta, \lambda)\right\}.$$

Here the coefficient $\delta k_o$ must vanish because of mass conservation. Coefficients $\delta h_o$, $\delta h_4$ and $\delta k_4$ are first order quantities in the ellipticity, $\varepsilon$. Also the deviations of the coefficients $h_2$ and $k_2$ from the Love numbers $h_2^o$ and $k_2^o$ in the spherical, non-rotating case are of first order in $\varepsilon$.

Since the deformed earth's surface remains an equipotential one, the potential increment at the earth's surface must be constant. This condition can be described to first order by the equation

(7-5) $$\begin{aligned} \text{const.} &= \left\{\varphi^P + \varphi^s\right\}_{r=r_o} - g\cdot u^{(n)} \\ &= \varphi^o\left\{\left(1 - \frac{4}{3}\varepsilon P_2\right)Y_{2m} + k_2\left(1 + 2\varepsilon P_2\right)Y_{2m} + \delta k_4 Y_{4m} - \right. \\ &\quad \left. - \left(1 - \frac{2}{3}m^* + \frac{2}{3}\beta P_2\right)\left(\delta h_o Y_{Om} + h_2 Y_{2m} + \delta h_4 Y_{4m}\right)\right\} \end{aligned}$$

where

$$g(\vartheta) = 1 - \frac{2}{3}m^* + \frac{2}{3}\beta P_2(\cos\vartheta)$$

is the gravity at the elliptical earth's surface, and the dimensionless parameters $m^*$ and $\beta$ are given by

$$m^* = \frac{\Omega^2 a^3}{GM}, \quad \beta = \frac{5}{2} m^* - \varepsilon .$$

The right hand of equation (7-5) can be expanded in terms of fully normalized spherical harmonics $Y_{nm}$ ( $n = 0, 2, 4$ ). The vanishing of the coefficients of the harmonics with non-zero degrees yields, to the first order, the following two relations:

$$(7\text{-}6) \quad \begin{cases} 1 + k_2 - h_2 = -\frac{2}{3} m^* h_2^o - \frac{1}{7} \alpha ( m^2 - 2 ) \\ \delta k_4 - \delta h_4 = \frac{1}{70} \alpha \sqrt{5 ( m^2 - 16 )( m^2 - 9 )} \end{cases}$$

with

$$(7\text{-}7) \quad \alpha = \frac{4}{3} \varepsilon - 2 \varepsilon k_2^o + \frac{2}{3} \beta h_2^o .$$

On the other hand, the deformation-induced Eulerian potential is caused only by the displacements at the earth's surface because there is no density redistribution in the earth's interior. To the first order, this part of the Eulerian potential is related to the vertical displacement at the earth's surface by the surface integral

$$(7\text{-}8) \quad \varphi^s( r, \vartheta, \lambda ) = \frac{3}{4\pi} \oiint_{\Sigma} \frac{u^{(n)} d\sigma}{| \mathbf{r} - \mathbf{r}_o |} .$$

The surface element $d\sigma$ at the elliptical surface is given by

$$d\sigma = r_o^2 \sin\vartheta_o d\vartheta_o d\lambda_o \sqrt{1 + \left( \frac{dr_o}{r_o d\vartheta_o} \right)^2} \approx r_o^2 \sin\vartheta_o d\vartheta_o d\lambda_o .$$

By substituting expression (7-3) for displacement $u^{(n)}$ and expanding the inverse distance $1 / | \mathbf{r} - \mathbf{r}_o |$ in terms of spherical harmonics, the deformation-induced Eulerian potential is given by

$$\begin{aligned} \varphi^s ( r, \vartheta, \lambda ) = & \frac{1}{r} 3 \left( \delta h_o - \frac{4}{3} \frac{1}{\sqrt{5}} \varepsilon h_2^o \delta_{m,o} \right) Y_{om}( \vartheta, \lambda ) + \\ & + \frac{1}{r^3} \frac{3}{5} h_2 \left( 1 + \frac{8}{21} ( m^2 - 2 ) \varepsilon \right) Y_{2m}( \vartheta, \lambda ) + \\ & + \frac{1}{r^5} \frac{1}{3} \left( \delta h_4 - \frac{2}{35} \sqrt{5 ( m^2 - 16 )( m^2 - 9 )} \; \varepsilon h_2^o \right) Y_{4m}( \vartheta, \lambda ). \end{aligned}$$

Comparing this equation with equation (7-4) yields the following three relations

$$
(7\text{-}9)\quad \begin{cases} \delta k_o = 3\left( \delta h_o - \frac{4}{3}\frac{1}{\sqrt{5}}\, \varepsilon h_2^o \delta_{m,o} \right) = 0 \\ k_2 = \frac{3}{5} h_2 \left( 1 + \frac{8}{21}( m^2 - 2 ) \varepsilon \right) \\ \delta k_4 = \frac{1}{3}\left( \delta h_4 - \frac{2}{35}\sqrt{5( m^2 - 16 )( m^2 - 9 )}\, \varepsilon h_2^o \right) \end{cases}
$$

Noting that for the homogeneous fluid earth, $h_2^o = 2.5$, $k_2^o = 1.5$, $m^* = 0.8\varepsilon$ and $\beta = \varepsilon$ ( $= 1/233.15$ ) and connecting equations (7-9) with (7-6), we obtain the following analytical solutions:

$$
(7\text{-}10)\quad \begin{cases} \delta h_o = \frac{2}{3}\sqrt{5}\; \varepsilon\, \delta_{m,o} \\ \delta h_2 = h_2 - h_2^o = \frac{10}{21}( 3m^2 + 1 )\varepsilon \\ \delta h_4 = -\frac{\varepsilon}{14}\sqrt{5( m^2 - 16 )( m^2 - 9 )} \\ \delta k_o = 0 \\ \delta k_2 = k_2 - k_2^o = \frac{2}{7}( 5m^2 - 3 )\varepsilon \\ \delta k_4 = \delta h_4 . \end{cases}
$$

By using the perturbation method, all these coefficients have been determined numerically by integrals given in Appendix C. As parameter perturbations there are

(i) gravity field perturbation $\delta V = \frac{1}{3}\left( \Omega^2 r^2 - 2\varepsilon g^o(r) r P_2( \cos\vartheta ) \right)$;

(ii) hydrostatic pressure perturbation $\nabla \delta P = \rho^o\, \nabla \delta_r V$ : and

(iii) surface undulation $\delta h( a ) = -\frac{2}{3}\varepsilon a P_2( \cos\vartheta )$.

In Table 1. we give both the analytical ( upper line ) and the numerical results ( lower line ) for comparison.

The agreement between the numerical and analytical results is almost perfect. this is due to the fact that, in this special case, the volumetric parameter perturbations $\delta V$ and $\delta P$ have in fact no contribution to the solution. The entire effect is attributed to the elliptical surface undulation $\delta h( a )$, and there is no progressive numerical error in the evaluation for this effect.

**Table 1.** Statical tidal deformations on an elliptical, rotating, incompressible, homogeneous, and fluid earth. Comparison between the analytical (upper line) and numerical results (lower line). All values are multiplied by a factor of 1000.

| m | $\delta h_o$ | $\delta h_2$ | $\delta h_4$ | $\delta k_2$ | $\delta k_4$ |
|---|---|---|---|---|---|
| | | ( * 1000 ) | | | |
| 0 | 6.394 | 2.042 | -8.220 | -3.676 | -8.220 |
| | 6.394 | 2.040 | -8.220 | -3.678 | -8.220 |
| 1 | —— | 8.170 | -7.504 | 2.451 | -7.504 |
| | —— | 8.169 | -7.504 | 2.451 | -7.504 |
| 2 | —— | 26.55 | -5.306 | 20.84 | -5.306 |
| | —— | 26.56 | -5.306 | 20.84 | -5.306 |

(b) Tidal deformations on an elliptical, rotating, incompressible, homogeneous and elastic earth

Theoretical studies of this problem go back to the beginning of this century. Love (1911) gave the entire expressions for both the influence of inertia due to the earth's rotation as well as that of the elliptical shape. Unfortunately, in his considering both effects separately, there is a violation of the presupposition of the hydrostatic equilibrium in the undeformed state. Furthermore, it is not easy to understand to which surface level his results should be related. In this section, we will not deduce the entire analytical solutions, but instead will consider a check for some partial relations of tidal parameters which can easily be obtained.

Since the third condition mentioned in the previous section, i.e., no density redistribution during the tidal motion in the earth's interior, is still valid in the present case, from equation (7-9) we get

$$(7\text{-}11)\quad \begin{cases} \delta h_o = \dfrac{4}{3}\dfrac{1}{\sqrt{5}}\,\varepsilon h_2^o\,\delta_{m,0} \\ \delta k_2 - \dfrac{3}{5}\delta h_2 = \dfrac{8}{35}\,(m^2-2)\,\varepsilon h_2^o \\ \delta k_4 - \dfrac{1}{3}\,\delta h_4 = -\dfrac{2}{105}\sqrt{5(m^2-16)(m^2-9)}\;\varepsilon h_2^o \end{cases}$$

In Table 2 the analytical (upper line) and the numerical results (lower line) are compared. The deviations of the numerical results from the analytical ones are on the level of one per mill. These errors, especially in the term of the zero degree for the zonal tide, are most probably caused by the fact that in our numerical procedure the incompressibility is imitated by a very large bulk modulus ( ≈ 100 times the shear modulus).

**Table 2.** Tidal deformations on an elliptical, rotating, incompressible, homogeneous, and elastic earth. Comparison between the analytical (upper line) and the numerical results (lower line). All values are multiplied by a factor of 1000.

| m | $\delta h_o / h_2^o$ | $\left(\delta k_2 - \frac{3}{5}\delta h_2\right)/h_2^o$ | $\left(\delta k_4 - \frac{1}{3}\delta h_4\right)/h_2^o$ |
|---|---|---|---|
| | | ( * 1000 ) | |
| 0 | 2.558<br>2.519 | -1.961<br>-1.965 | -2.192<br>-2.193 |
| 1 | —<br>— | -0.980<br>-0.983 | -2.001<br>-2.002 |
| 2 | —<br>— | -1.961<br>-1.963 | -1.415<br>-1.416 |

(c) Statical tidal deformations on an elliptical, rotating, compressible, stratified and fluid earth

In order to check the validity of the numerical evaluation for contributions of various volumetric parameter perturbations and internal boundary undulations, we consider here the tidal deformations on an elliptical, rotating, compressible, stratified, and fluid earth. As the starting earth model, we take the spherically symmetrical density distribution from PREM (Dziewonski & Anderson, 1981). The bulk modulus is determined by the so-called Adams-Williamson condition for stability (Longman, 1963).

As in the case of Section (b), we can derive only some relations between the tidal parameters. Using the condition of the equipotential earth's surface, we obtain from equations (7-6)

$$\text{(7-12)} \quad \begin{cases} \delta k_2 - \delta h_2 = -\frac{2}{3} m^* h_2^o - \frac{1}{7}\alpha(m^2 - 2) \\ \delta k_4 - \delta h_4 = \frac{1}{70}\alpha\sqrt{5(m^2 - 16)(m^2 - 9)} \end{cases}$$

where $\alpha$ is given by (7-7). The hydrostatic Love numbers for earth model PREM are

$h_2^o = 1.94197, \quad k_2^o = 0.94197.$

based on our calculations. If we take m = 0.0034312, then to first order we have $\varepsilon = 1/300.15$, $\beta = 0.0052318$. Using these parameters, in Table 3 we give the comparison between the analytical (upper line) and the numerical results (lower line) for the relations (7-12).

**Table 3.** Statical tidal deformations on an elliptical, rotating, compressible, stratified and fluid earth (PREM). Comparison between the analytical (upper line) and numerical results (lower line). All values are multiplied by a factor of 1000.

| m | $\delta k_2 - \delta h_2$ | $\delta k_4 - \delta h_4$ |
|---|---|---|
| | ( * 1000 ) | |
| 0 | -3.031<br>-3.081 | 1.893<br>1.882 |
| 1 | -3.737<br>-3.761 | 1.728<br>1.718 |
| 2 | -5.853<br>-5.803 | 1.222<br>1.215 |

Here again the numerical results have an accuracy on the level of one per mill as before.

7.3 Starting model for the realistic earth

In the following calculations, we take the spherically symmetric, non-rotating, elastic and isotropic (SNREI) earth model as the undisturbed starting earth model. The equilibrium values of the rheological parameters $\lambda^o(r)$, $\mu^o(r)$ and $\rho^o(r)$ are taken from PREM.

Although PREM is the most modern seismic reference earth model, it is not very appropriate for tidal calculations since it includes a fluid ocean surface. For our purpose, this thin fluid surface layer ( ~ 3 km) will be replaced by the solid crust layer. This small modification affects the gravity field, but the effect is below one per mill and is negligible.

From a variety of geophysical observations such as the damping of seismic waves and free oscillations, it is known that the earth's mantle deviates from ideal elasticity. The shear modulus is frequency dependent while the frequency dependence of the bulk modulus is not significant. Like all other seismic reference earth models, PREM is valid for some seismic reference frequency. In order to investigate the dispersion of tidal parameters due to inelasticity in the earth's mantle, it is appropriate to choose the unrelaxed (very high frequency) moduli for the starting model. There exist, at present, many rheological models to estimate the degree of anelasticity of the shear modulus in the seismic band, for example, the so-called absorption band model with a flat or slightly inclined frequency dependence of the quality factor Q over an arbitrary finite frequency range (Liu et al., 1976; Kanamori & Anderson, 1977; Anderson & Minster, 1979; Minster, 1980; Zschau, 1986). Recently, based on a large set of data from the damping of seismic waves and free oscillations, dispersion of tidal parameters, quality factor of the Chandler wobble, post glacial rebound data as well as laboratory experiments, a new rheological model was given by Zschau (see Zschau & Wang, 1985). This model supposes a generalized Maxwell rheology with a cut Gaussian distribution of the logarithmic stress relaxation times and accounts for depth and frequency dependence of the shear modulus in the mantle in a broad frequency range. Dehant & Zschau (1989) pointed out that the best choice between all possible models, presently, is Zschau's model. In this thesis I will use this rheological model to determine the unrelaxed PREM shear modulus and its visco-elastic dispersion. The formulae for determining the depth and frequency dependence of the shear modulus according to this model are given in Appendix D and in Section 7.5.

## 7.4 Comparison with Wahr's theory of earth tides

Based on the previous work of Smith (1974), Wahr (1979) computed the tidal deformations of an elliptical, rotating, elastic and oceanless earth by integrating a large set of equations of motion using the generalized spherical harmonic functions (Edmonds, 1960; Phinney & Burridge, 1973). Unlike earlier theories, Wahr's theory is considerably more complicated and complete in the sense that it accounts for the effect of rotation and takes into account elliptical stratification throughout the whole earth. The resulting effects are of the order of 1%. It is at present difficult to verify experimentally these small differences between Wahr's model and a non-rotating spherical model because of other insufficiently modelled influences such as the effect of ocean tidal loading, calibration errors etc. In spite of this, Wahr's model of earth tides has been used recently as the a priori standard in modelling satellite tracking data (Marsh et al., 1988).

In order to possibly provide a check for Wahr's theory, here we use the perturbation method to compute an independent solution of the same problem. To get easily comparable results, we use the same representations for all interesting tidal observables, limited to the tides of the second degree. For convenience, we remember the formulae given in Wahr (1979):

Surface displacement:

$$(7\text{-}18)\qquad \mathbf{u} = \mathbf{e}_r h_o\Big[\big(1 + yP_2\big)Y_{2m} + h_+ Y_{4m} + h_- Y_{0m}\Big] + $$
$$+ l_o\Big[\big(1 + zP_2\big)\nabla_1 Y_{2m} + l_+ \nabla_1 Y_{4m} + $$
$$+ \mathbf{e}_r \times \big(w_+ \nabla_1 Y_{3m} + w_- \nabla_1 Y_{1m}\big)\Big].$$

Eulerian potential in free space:

$$(7\text{-}19)\qquad \varphi^s = k_o\Big[\frac{1}{r^3} Y_{2m} + k_+ \frac{1}{r^5} Y_{4m}\Big].$$

Gravity:

$$(7\text{-}20)\qquad g_t^{(n)} = -2\Big[G_o Y_{2m} + G_+ Y_{4m} + G_- Y_{0m}\Big].$$

Tilt:

$$(7\text{-}21a)\qquad t_t^{(s)} = T_o^s \frac{\partial}{\partial\vartheta} Y_{2m} + T_1^s \frac{\partial}{\partial\vartheta} Y_{4m} + T_3^s \frac{m}{\sin\vartheta} Y_{3m} + T_4^s \frac{m}{\sin\vartheta} Y_{1m},$$

$$(7\text{-}21b)\qquad t_t^{(e)} = T_o^e \frac{m}{\sin\vartheta} Y_{2m} + T_1^e \frac{m}{\sin\vartheta} Y_{4m} + T_3^e \frac{\partial}{\partial\vartheta} Y_{3m} + T_4^e \frac{\partial}{\partial\vartheta} Y_{1m}.$$

Here the unit system defined in 7.1 has been used, and without loss of generality, a unit amplitude for the tide-generating potential has been supposed.

In Table 4, we compile both, the results of Wahr (1979) for the starting earth model PEM-C and our results for the starting earth model PREM. From the comparison we find a good agreement in the surface displacement, except for the toroidal components. The important discrepancies are found in the coefficients which describe the latitude dependence of the Love number k, the gravimetric factor and the tilt factor. For all the diurnal, semi-diurnal and long period tides, the latitude dependence of the Love number k and the gravimetric factor given in the present thesis is only half as large as that given by Wahr (1979). Additionally, in our results, coefficient sets $T^s$ and $T^e$ , which describe the two tilt components, are identical for both, the north-south and east-west directions while they are quite different in Wahr's results.

It should be noted that such large discrepancies between our present results and Wahr's results are neither due to the insufficient approximation in the perturbation method nor due to the different starting earth model used. In fact, in order to truncate the infinite coupled system of differential equations to a finite coupled system of, in general case, 22 equations, Wahr had to neglect all the apparently minimal quantities of second or higher order. Furthermore, this finite system was partially uncoupled by making further approximations. Although the comparison is made between our results based on the starting model PREM and Wahr's results based on the starting model PEM-C, there should be no significant numerical differences in the effect of rotation and ellipticity. Another difference is that here the theoretical tides are evaluated at a reference sphere with the mean earth's radius while in Wahr they are related to a reference sphere with the equatorial radius. This difference only results in a small effect of about 0.2% in parameters $h_o$, $l_o$, $k_o$ and in parameter sets G and T. For the same reference level, in Table 4 for these parameters an increase of 0.2% in Wahr's results or a decrease of 0.2% in our results should be introduced.

However, it may be theoretically verified that the coefficient sets for describing the two orthogonal tilt components should be identical, or, in other words, the two orthogonal tilt components expanded in terms of spherical harmonics, must take a coupled form. Using the previous notations, the tilt vector on the deformed earth's surface can be expressed as

$$\mathbf{t} = \mathbf{t}^o + \Delta\mathbf{t}$$

where $\mathbf{t}^o$ is the undisturbed tilt vector related to the spherical earth's surface, and $\Delta\mathbf{t}$ the so-called total variation of the tilt vector. In our unit system, we can express $\mathbf{t}^o$ as

$$\mathbf{t}^o = \gamma^o_{n_o} \nabla_1 Y_{n_o m_o}.$$

Here $\gamma^o_{n_o}$ is the undisturbed tilt factor for degree $n_o$. In general, vector $\Delta\mathbf{t}$ can be always expanded in terms of spherical harmonics:

$$\Delta\mathbf{t} = \sum_n \sum_m \left\{ a^m_n \mathbf{e}_r + b^m_n \nabla_1 + c^m_n \mathbf{e}_r \times \nabla_1 \right\} Y_{nm}.$$

Supposing the surface undulation of the spherical shape to be $\delta h(\vartheta, \lambda)$, then the surface unit vectors in the south direction and in the east direction are

$$\mathbf{e}_s = \mathbf{e}_\vartheta + \mathbf{e}_r \frac{\partial}{\partial\vartheta}\delta h \quad \text{and} \quad \mathbf{e}_e = \mathbf{e}_\lambda + \mathbf{e}_r \frac{1}{\sin\vartheta}\frac{\partial}{\partial\lambda}\delta h,$$

respectively. Note that since $\Delta\mathbf{t}$ is of first order, the tilt component is therefore given to first order by

$$t^{(s)} = \mathbf{e}_s \cdot \mathbf{t} = \gamma^{o}_{n_o} \frac{\partial}{\partial\vartheta} Y_{n_o m_o} + \sum_n\sum_m \left\{ b^m_n \frac{\partial}{\partial\vartheta} Y_{nm} - c^m_n \frac{1}{\sin\vartheta}\frac{\partial}{\partial\lambda} Y_{nm} \right\}$$

in the south direction, and

$$t^{(e)} = \mathbf{e}_e \cdot \mathbf{t}$$

$$= \gamma^{o}_{n_o} \frac{1}{\sin\vartheta}\frac{\partial}{\partial\lambda} Y_{n_o m_o} + \sum_n\sum_m \left\{ b^m_n \frac{1}{\sin\vartheta}\frac{\partial}{\partial\lambda} Y_{nm} + c^m_n \frac{\partial}{\partial\vartheta} Y_{nm} \right\}$$

in the east direction. Considering that

$$t_{nm} = b^m_n + \gamma^{o}_{n_o} \delta_{n,n_o} \delta_{m,m_o} \quad \text{and} \quad \omega_{nm} = c^m_n ,$$

the two tilt components are expressed in a coupled form by

$$\begin{cases} t^{(s)} = \sum_n\sum_m \left\{ t_{nm} \frac{\partial}{\partial\vartheta} Y_{nm} - \omega_{nm} \frac{1}{\sin\vartheta}\frac{\partial}{\partial\lambda} Y_{nm} \right\} \\ t^{(e)} = \sum_n\sum_m \left\{ t_{nm} \frac{1}{\sin\vartheta}\frac{\partial}{\partial\lambda} Y_{nm} + \omega_{nm} \frac{\partial}{\partial\vartheta} Y_{nm} \right\} . \end{cases}$$

It is thus verified that the two coefficient sets $T^s$ and $T^e$ given in Table 4 have to be the same. From the differences in these two coefficient sets of Wahr's results, we conclude that there are probably some mistakes in the formulation of Wahr's theory.

It has been mentioned in the previous chapter that Dehant (1987) redefined the gravimetric factor as the transfer function between the earth's tidal response measured by a gravimeter and the vertical component of the external tide-generating force, both related to the elliptical surface. This modification of the classical definition leads to a reduction of the latitude dependence of the gravimetric factor deduced from Wahr's results to half its size. Adopting Dehant's definition, the gravimetric factor, as deduced from our results has in the first order no more latitude dependence. It turns out that

$$\text{(7-22)} \qquad \delta_2 = \begin{cases} 1.1570, & \text{for long period tides;} \\ 1.1581, & \text{for diurnal tides;} \\ 1.1613, & \text{for semi-diurnal tides.} \end{cases}$$

These results show a slight increase of 0.2%, 0.4% and 0.3% compared to the $\delta_2$-value averaged over the elliptical surface as obtained by Dehant & Zschau (1989) for Mf, O1 and M2, respectively. The earth model adopted by them is the same as the one used here. Compared to the observed values $\delta_2$ = 1.1611 for O1 and 1.1707 for M2 (Dehant & Ducarme, 1987), the present results fit the observations very well. The discrepancy is only about 0.3% and 0.8%, respectively.

**Table 4.** Tidal deformations on an elliptical, rotating, elastic, and ocean-less earth. Comparison between the results of Wahr (1979) (left column) and of the present thesis (right column).

| | m = 0 long period | | m = 1 diurnal | | m = 2 semi-diurnal | |
|---|---|---|---|---|---|---|
| $h_o$ | 0.606 | 0.6050 | 0.603 | 0.6059 | 0.609 | 0.6079 |
| $h_+$ | 0.000 | -0.0006 | 0.000 | -0.0005 | 0.000 | -0.0004 |
| $h_-$ | 0.000 | -0.0006 | —— | —— | —— | —— |
| $y$ | 0.001 | 0.0010 | 0.001 | 0.0010 | 0.001 | 0.0010 |
| $l_o$ | 0.084 | 0.0839 | 0.084 | 0.0842 | 0.085 | 0.0850 |
| $l_+$ | -0.002 | -0.0026 | -0.002 | -0.0023 | -0.001 | -0.0010 |
| $z$ | 0.014 | 0.0138 | 0.014 | 0.0138 | 0.014 | 0.0138 |
| $w_+$ | 0.000 | 0.0000 | 0.000 | 0.0043 | 0.001 | 0.0068 |
| $w_-$ | 0.000 | 0.0000 | 0.000 | 0.0000 | —— | —— |
| $k_o$ | 0.299 | 0.2986 | 0.298 | 0.2991 | 0.302 | 0.3009 |
| $k_+$ | -0.005 | -0.0029 | -0.005 | -0.0026 | -0.003 | -0.0018 |
| $G_o$ | 1.155 | 1.1574 | 1.152 | 1.1583 | 1.160 | 1.1609 |
| $G_+$ | -0.007 | -0.0032 | -0.007 | -0.0029 | -0.005 | -0.0021 |
| $G_-$ | 0.005 | 0.0005 | —— | —— | —— | —— |
| $T_o^s$ | 0.698 | 0.6945 | 0.689 | 0.6962 | 0.693 | 0.7013 |
| $T_1^s$ | -0.001 | -0.0024 | -0.001 | -0.0022 | 0.000 | -0.0015 |
| $T_3^s$ | —— | —— | -0.001 | 0.0004 | -0.001 | 0.0006 |
| $T_4^s$ | —— | —— | 0.004 | -0.0019 | —— | —— |
| $T_o^e$ | —— | —— | 0.689 | 0.6962 | 0.689 | 0.7013 |
| $T_1^e$ | —— | —— | 0.000 | -0.0022 | 0.000 | -0.0015 |
| $T_3^e$ | 0.000 | 0.0000 | -0.001 | 0.0004 | -0.002 | 0.0006 |
| $T_4^e$ | 0.000 | 0.0000 | 0.002 | -0.0019 | —— | —— |

### 7.5 Visco-elastic dispersion of Love numbers

It is known that the frequency dependence of shear modulus is modulated by the chemical composition, structure, phase transitions, temperature and pressure. According to Zschau's relaxation theory given in Appendix D, this effect can be empirically described by a single parameter

$$(7\text{-}23) \qquad z = \frac{\rho_o T}{\Phi}$$

where $\rho_o$ is the ambient density of the material at the normal condition, T the temperature, and $\Phi$ the seismic parameter.

The distribution of this parameter in the earth's mantle has been determined based on his temperature model (personal communication) and can be found in Table 8. In general, we may suppose

$$(7\text{-}24) \qquad \mu(z, \omega) = \mu_o(z) \cdot F(z, \omega).$$

Here $\omega$ is the frequency, $\mu_o$ is called the unrelaxed or very high frequency modulus. F is the relaxation factor and given in Table 5.

If the relaxed modulus, $\mu(z, \omega)$, is known for a given seismic reference frequency $\omega_o$, then the unrelaxed modulus is determined by

$$(7\text{-}25) \qquad \mu_o(z) = \frac{\mu(z, \omega_o)}{F(z, \omega_o)},$$

and

$$(7\text{-}26) \qquad \mu(z, \omega) = \mu(z, \omega_o)\frac{F(z, \omega)}{F(z, \omega_o)}.$$

With the above formulae, we determine the relaxed shear modulus in the tidal band. It is, in general, a complex function. Using the correspondence principle, the tidal deformations in the frequency domain can be calculated as in the elastic case. The Love numbers as well as all other tidal parameters are thus complex. They can be determined by integrating the equations of motion described in Appendix A in the complex plane (see also Zschau, 1979). As an indirect method, this can also be done by our perturbation method since in the tidal band, the visco-elastic deviation of the shear modulus from its elastic value is small. Another purpose of doing this is to check the validity of the perturbation method itself or, in other words, to see what does the expression "small" mean in the perturbation method?

**Table 5.** Relaxation factor in the earth's mantle for the seismic reference period 200s, the semi-diurnal tide ($M_2$), the diurnal tide ($O_1$), the fortnightly tide ($M_f$), the semiannual tide ($S_{sa}$) and the Chandler wobble (CW). The determination is based on Zschau's relaxation theory (Appendix D) and his rheological model (Table 8).

| depth | relaxation factor F ( real / imaginary part ) | | | | | |
|---|---|---|---|---|---|---|
| (km) | 200s | $M_2$ | $O_1$ | $M_f$ | $S_{sa}$ | CW |
| 24- 60 | 0.9998<br>0.0004 | 0.9968<br>0.0009 | 0.9964<br>0.0010 | 0.9944<br>0.0016 | 0.9912<br>0.0024 | 0.9898<br>0.0028 |
| 60- 80 | 0.9932<br>0.0026 | 0.9768<br>0.0085 | 0.9726<br>0.0100 | 0.9519<br>0.0172 | 0.9154<br>0.0298 | 0.8980<br>0.0357 |
| 80- 150 | 0.9815<br>0.0079 | 0.9261<br>0.0306 | 0.9112<br>0.0366 | 0.8335<br>0.0670 | 0.6877<br>0.1213 | 0.6156<br>0.1459 |
| 150- 220 | 0.9631<br>0.0172 | 0.8367<br>0.0726 | 0.8013<br>0.0874 | 0.6112<br>0.1605 | 0.2530<br>0.2204 | 0.1148<br>0.1798 |
| 220- 310 | 0.9862<br>0.0057 | 0.9472<br>0.0212 | 0.9369<br>0.0251 | 0.8839<br>0.0453 | 0.7861<br>0.0815 | 0.7383<br>0.0987 |
| 310- 400 | 0.9891<br>0.0044 | 0.9598<br>0.0157 | 0.9522<br>0.0185 | 0.9133<br>0.0330 | 0.8426<br>0.0586 | 0.8083<br>0.0708 |
| 400- 500 | 0.9897<br>0.0041 | 0.9624<br>0.0145 | 0.9554<br>0.0172 | 0.9194<br>0.0304 | 0.8543<br>0.0539 | 0.8227<br>0.0651 |
| 500- 670 | 0.9903<br>0.0038 | 0.9649<br>0.0135 | 0.9584<br>0.0159 | 0.9251<br>0.0281 | 0.8651<br>0.0496 | 0.8361<br>0.0598 |
| 670- 771 | 0.9854<br>0.0061 | 0.9435<br>0.0228 | 0.9325<br>0.0271 | 0.8752<br>0.0490 | 09692<br>0.0884 | 0.7173<br>0.1070 |
| 771-1271 | 0.9891<br>0.0044 | 0.9598<br>0.0157 | 0.9522<br>0.0185 | 0.9133<br>0.0330 | 0.8426<br>0.0586 | 0.8083<br>0.0708 |
| 1271-1771 | 0.9944<br>0.0021 | 0.9811<br>0.0067 | 0.9778<br>0.0079 | 0.9616<br>0.0135 | 0.9332<br>0.0230 | 0.9198<br>0.0275 |
| 1771-2271 | 0.9958<br>0.0015 | 0.9867<br>0.0046 | 0.9845<br>0.0053 | 0.9736<br>0.0089 | 0.9551<br>0.0149 | 0.9464<br>0.0176 |
| 2271-2741 | 0.9967<br>0.0011 | 0.9900<br>0.0033 | 0.9883<br>0.0038 | 0.9805<br>0.0063 | 0.9674<br>0.0104 | 0.9613<br>0.0123 |
| 2741-2891 | 0.9961<br>0.0014 | 0.9876<br>0.0042 | 0.9855<br>0.0049 | 0.9755<br>0.0081 | 0.9585<br>0.0136 | 0.9506<br>0.0161 |

The elastic Love numbers for the starting model PREM are

(7-27) $h_2 = 0.60548, \quad l_2 = 0.08440, \quad k_2 = 0.29921$

Their visco-elastic dispersions calculated by both, the direct integration method and the perturbation method are given in Table 6.

**Table 6.** Visco-elastical dispersion of Love numbers calculated by both the direct integration method (left column) and the perturbation method (right column). For each tidal period, their real and imaginary part are expressed in the upper and lower line, respectively.

| period | $\delta h_2$ | | $\delta l_2$ | | $\delta k_2$ | |
|---|---|---|---|---|---|---|
| $M_2$ | 0.00572 | 0.00566 | 0.00207 | 0.00201 | 0.00347 | 0.00343 |
| | -0.00210 | -0.00204 | -0.00082 | -0.00076 | -0.00130 | -0.00126 |
| $N_2$ | 0.00574 | 0.00570 | 0.00208 | 0.00202 | 0.00349 | 0.00345 |
| | -0.00210 | -0.00205 | -0.00082 | -0.00077 | -0.00131 | -0.00126 |
| $O_1$ | 0.00674 | 0.00666 | 0.00246 | 0.00238 | 0.00410 | 0.00404 |
| | -0.00246 | -0.00239 | -0.00098 | -0.00090 | -0.00154 | -0.00148 |
| $M_f$ | 0.01187 | 0.01160 | 0.00453 | 0.00427 | 0.00733 | 0.00712 |
| | -0.00436 | -0.00410 | -0.00190 | -0.00159 | -0.00276 | -0.00256 |
| $M_m$ | 0.01388 | 0.01350 | 0.00537 | 0.00501 | 0.00860 | 0.00831 |
| | -0.00511 | -0.00475 | -0.00234 | -0.00185 | -0.00326 | -0.00297 |
| $S_{sa}$ | 0.02117 | 0.02027 | 0.00848 | 0.00766 | 0.01331 | 0.01256 |
| | -0.00798 | -0.00706 | -0.00460 | -0.00274 | -0.00511 | -0.00438 |
| $S_a$ | 0.02473 | 0.02350 | 0.00998 | 0.00892 | 0.01563 | 0.01458 |
| | -0.00952 | -0.00814 | -0.00647 | -0.00313 | -0.00603 | -0.00502 |
| CW | 0.02574 | 0.02439 | 0.01040 | 0.00927 | 0.01669 | 0.01513 |
| | -0.00997 | -0.00844 | -0.00714 | -0.00324 | -0.00629 | -0.00519 |

From this table, we can conclude that the mantle inelasticity produces an increase of about 1.0% for h, 2.4% for l and 1.1% for k for the semi-diurnal tides, 1.1% for h, 2.8% for l and 1.4% for k for the diurnal tides and 2.0% for h, 5.4% for l and 2.4% for k for the zonal tide Mf. For the Chandler wobble period, this increase reaches up to 4.3% for h, 12% for l and 5.4% for k. Small phase shifts of about 0.2° for h, 0.5° for l and 0.2° for k are to be expected for the semidiurnal tides.

Based on our model calculations, in Table 5, the shear modulus in the upper mantle can be relaxed by about 5% for the main tidal period, 10% for the monthly period and 20% for the yearly period. In the asthenosphere, this relaxation can reach as much as 20%, 40% and 75%, respectively. From the comparison between the results of the direct integration method and the perturbation method, we may conclude that the perturbation method has an accuracy on the one per cent level if the maximum rheological parameter perturbations are not much more than 10 per cent. The Love numbers $h_2$ and $k_2$ determined by the perturbation method agree very well with the results of the direct integration method up to the yearly period. For the Love number $l_2$ the discrepancy becomes significant if the period is longer than about one month. It is because $l_2$ is very sensitive to the relaxation in the asthenosphere.

7.6 Tidal tomography and mantle heterogeneities

Seismic tomography has clearly shown that the earth's mantle is laterally inhomogeneous (Masters et al., 1982; Nakanishi & Anderson, 1982; Dziewonski et al., 1977). A global three-dimensional model has recently been constructed for the S velocity in the upper mantle, expanded up to degree and order 8 in spherical harmonics, and for the P velocity in the lower mantle, up to degree and order 6. (Dziewonski, 1984; Woodhouse & Dziewonski, 1984). The predicted heterogeneities are of the order of ± 1-1.5% in the lower mantle and ±1- 2.5% in the upper mantle. The major features of continental shields and ocean ridges are reflected by large velocity anomalies of up to the order of ± 8% in the depth interval from the Moho-discontinuity to 250 km.

As a consequence, the tidal parameters should be dependent not only upon the latitude but also upon the longitude and, differently from the effect due to rotation and ellipticity, additional phase shifts should be caused by the axial asymmetry of the rheological parameter distribution. To estimate the magnitude of this influence, in particular on the gravimetric factor, Molodenskij (1977, 1980) has applied the perturbation method. For an unrealistic model, in which the seismic velocities in the whole mantle under the continent are uniformly larger by 5% than those under the ocean, the resulting effect in the gravimetric factor is of the order of up to 0.75% (Molodenskij & Kramer, 1980). He did not, however, take into account the effect of interior boundary undulations, the density heterogeneities and the resulting perturbations of the hydrostatic pre-stress equilibrium, assuming these effects to be small.

Compared to the results of the seismic tomography, lateral heterogeneities of 5% throughout the mantle as used in Molodenskij's continent-ocean model, are extremely large. However, if the lateral heterogeneities are partly due to temperature differences, they might show dispersion and could be considerably stronger in the tidal band than at seismic periods (Wang & Zschau, 1986).

Molodenskij's assumption that the effect due to lateral density heterogeneities can be neglected, is based on the fact that these density heterogeneities are rather small in the earth's interior as estimated from the gravity anomalies, compared to the heterogeneities in the moduli of elasticity. However, the influence of a rheological parameter perturbation on earth tides depends not only upon its quantity but also upon the sensitivity to this perturbation. Therefore, the validity of this assumption should be examined.

Before we calculate the influences of mantle heterogeneities on earth tides for a realistic three dimensional model, it is very important to investigate the sensitivities of different tidal parameters to the depth and wavelength of different earth parameter perturbations.

## 7.7 Sensitivity of tidal parameters to mantle heterogeneities

For the investigation of sensitivities we consider the three Love numbers h, k and l and the gravimeter factor $\delta$ in the semi-diurnal band. As has been mentioned, the Love number l is anisotropic. The lateral heterogeneities are those in the Lamė constants, $\lambda$ and $\mu$, and in the density, $\rho$, in the upper and lower mantle, and the lateral undulations at the core-mantle boundary, and at the Moho-discontinuity. To investigate the sensitivity to the spatial wavelength, all these perturbations are described by single zonal harmonics of different degrees (2, 5 and 8). The reason for the choice of zonal harmonics is the remaining of axial symmetry, so that the effect in the tidal parameters depends only upon latitude and thus it can easily be graphically illustrated.

It has been shown that deviations from the spherically symmetrical density distribution or boundary undulations at a boundary with density discontinuity cause variations not only in the earth's gravity field but also in the initial pre-stress equilibrium. The variation in the gravity field has been formulated in Chapter 4, and its influences on earth tides can be evaluated by the formulae given in Appendix C. Since the variation in the initial hydrostatic pre-stress is not uniquely determined by the perturbation of the density distribution and the interior boundary undulation, we in general cannot fully account for this indirect effect. However, the variation in the generalized hydrostatic pressure introduced in Chapter 2 can be uniquely determined (s. Chapter 4), and it should represent the main contribution of the variation in the initial hydrostatic pre-stress, if the earth is to first order in isostatic equilibrium. In this section, the latitude dependence of tidal parameters is due to a zonal lateral heterogeneity of the density distribution in the mantle and due to a zonal undulation at the core-mantle as well as at the Moho-discontinuity. This includes the indirect effects of the resulting variations in the earth's gravity field and in the generalized hydrostatic pressure.

Since the perturbation method is valid to first order, the variation in the tidal parameters is linearly related to the variation in the earth's model parameters. For simplicity, for each harmonic variation in the volumetric parameter $\delta\rho$, $\delta\lambda$ and $\delta\mu$ we choose a maximum relative deviation of 5% from the corresponding average value in the surroundings, and a maximum absolute deviation of 10 km for the boundary undulation at the core-mantle and at the Moho-discontinuity. The results can be linearly interpolated for similar parameter variations with any other amplitudes.

The results are illustrated in Fig. 1-5. As expected, the effects increase in most cases with the wavelength of the lateral heterogeneity in the earth's model parameters. For low degree harmonics, there is no general decrease with depth of the heterogeneous layer. Instead, the general characteristics of the curves are changing with depth in most cases. This is due to a varying sign of some kernel functions (K and G given in Appendix C) in the mantle.

The effect in the Love number h, as shown in Fig. 1, is weakly sensitive to $\delta\lambda$ in the whole mantle, $\delta\mu$ in the lower mantle and $\delta\rho$ in the upper mantle. It can be concluded that, when the error of the Love number h is on the per mill level, tidal data cannot enable us to detect lateral heterogeneities smaller than 5%. But with the same accuracy of the measurement we could resolve $\delta\mu$ of about 1% in the upper mantle, $\delta\rho$ of about 0.5% in the lower mantle, lateral undulations of about 1 km at the Moho-discontinuity or of about 4 km at the core-mantle boundary. This conclusion, however, is only valid for the very long wavelength of the lateral heterogeneities. The sensitivity of the effect in the Love number h for the short wavelengths is considerably weaker.

Similar characteristics can be found in Fig. 2 for the Love number k, except that its sensitivity to $\delta\mu$ in the upper mantle and to the undulations at the Moho-discontinuity is twice as strong as that of Love number h, and conversely to $\delta\rho$ in the lower mantle. The effect on the gravimetric factor, $\delta$, is systematically smaller than that on the Love numbers h and k. It can be concluded that the large-scale lateral heterogeneities in the earth's mantle, as estimated by seismic tomography, cannot be resolved by tidal gravity data if the accuracy of the measurement is not better than one per mill.

A large percentage effect is to be expected in the horizontal displacement as shown in Fig. 4 and 5. Different from the effect in Love number h and k and in the gravimetric factor $\delta$, the effect in both, the north-south and east-west component of Love number l does not decrease significantly with the degree of the spherical harmonics, and sometimes it even behaves oppositely. At the equator, where the horizontal displacement of a sectorial tide in the undisturbed spherically symmetric system has no component in the north-south direction, its percentage variation becomes divergent.

It should be noted that different lateral heterogeneities may influence the tidal parameters with opposite effects, and as a consequence the total effect can be strongly reduced when they appear together with certain proportion. Furthermore, the similar behavior of Love number h and k cancels to a certain extent the effect in the tidal gravity because these two Love numbers enter in the gravimetric factor with different signs.

A general description for Fig. 1-5: — *Sensitivities of tidal parameters in the semi-diurnal band to the depth and wavelength of mantle heterogeneities and internal boundary undulations* — The considered earth's parameter perturbations are denoted by

**LA** for deviation of the first Lamé const. (max. 5%),

**MU** for deviation of the second Lamé const. (max. 5%),

**RO** for deviation of the density (max. 5%), and

**BU** for boundary undulation (max. 10 km).

Their distributions are described by the single zonal harmonics of degree

**l = 2**: solid line,

**l = 5**: broken line, and

**l = 8**: dotted line,

and they are placed in the following regions

**UM**: upper mantle,

**LM**: lower mantle,

**MH**: Moho-discontinuity, and

**CM**: core-mantle boundary, respectively.

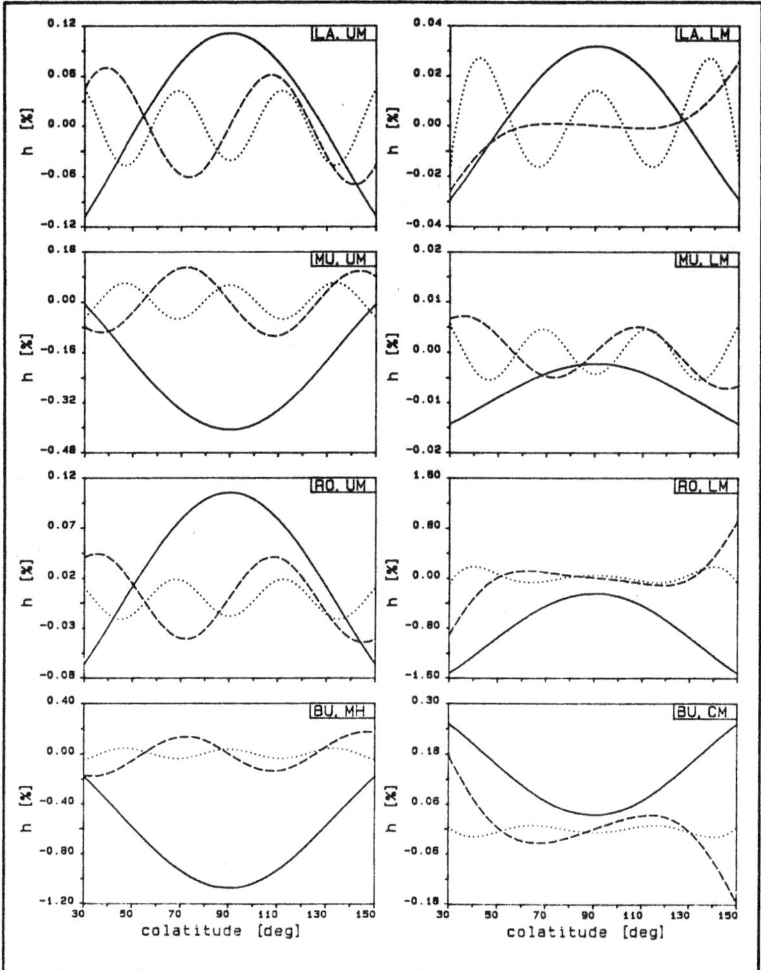


**Figure 1.** Latitude dependence of Love number h for a sectorial tide due to different zonal mantle heterogeneities as well as undulations at the core-mantle boundary and the Moho-discontinuity. For further details see the general description on the foregoing page.

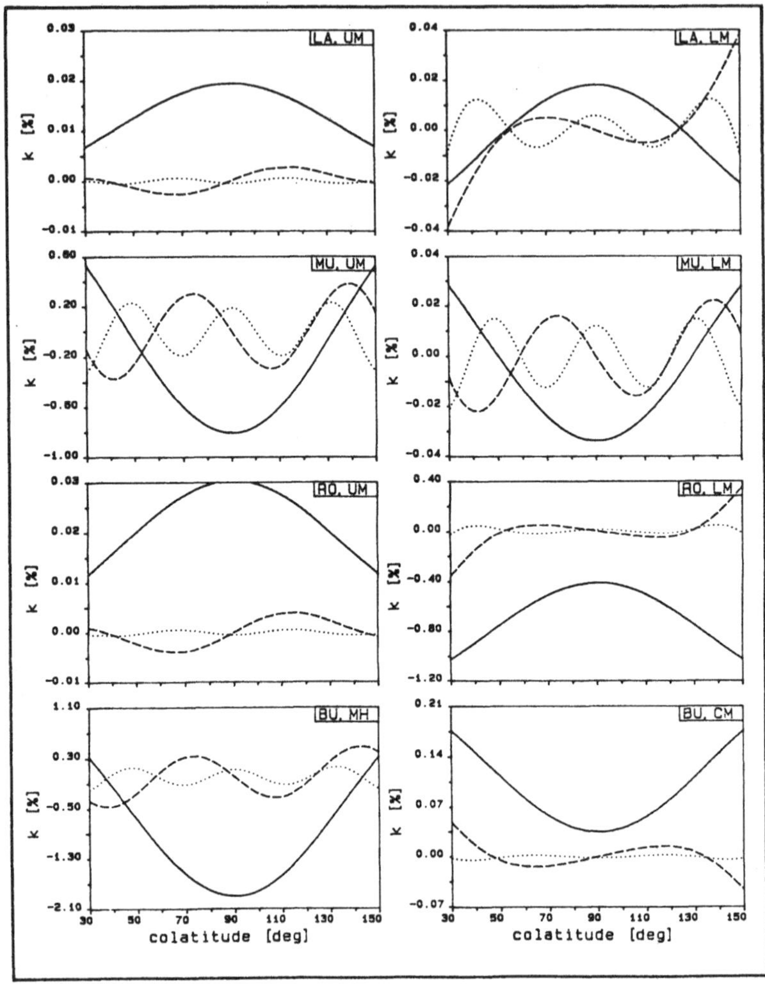


**Figure 2.** Same as Figure 1, but for Love number k.

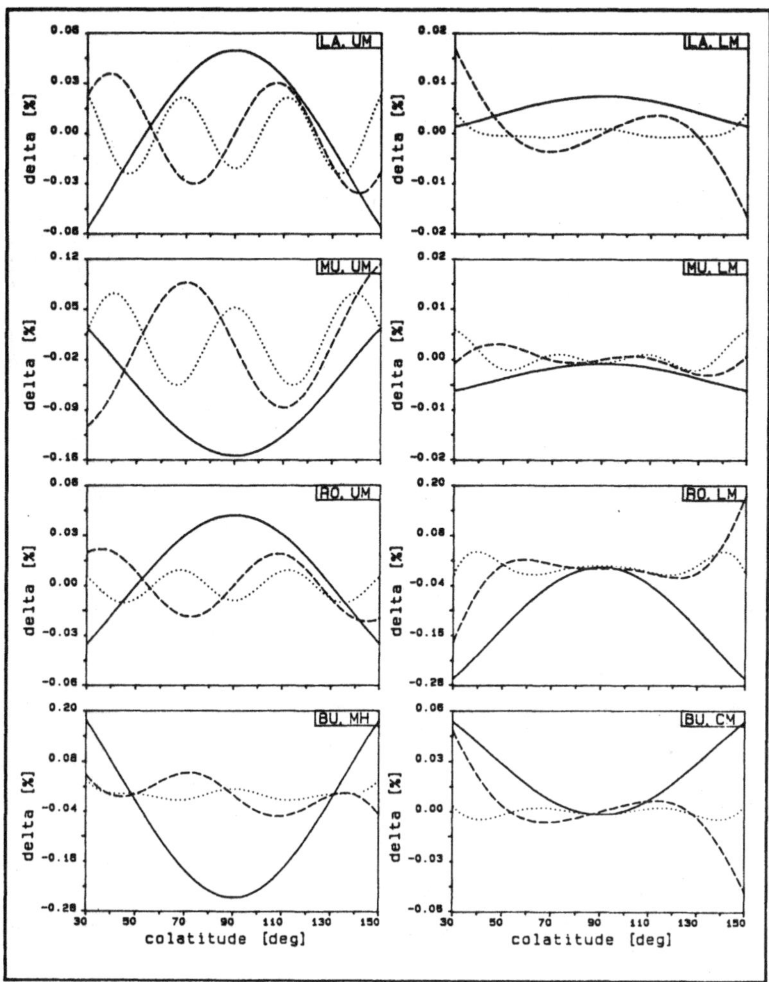


**Figure 3.** Same as Figure 1. but for the gravimetric factor $\delta$.

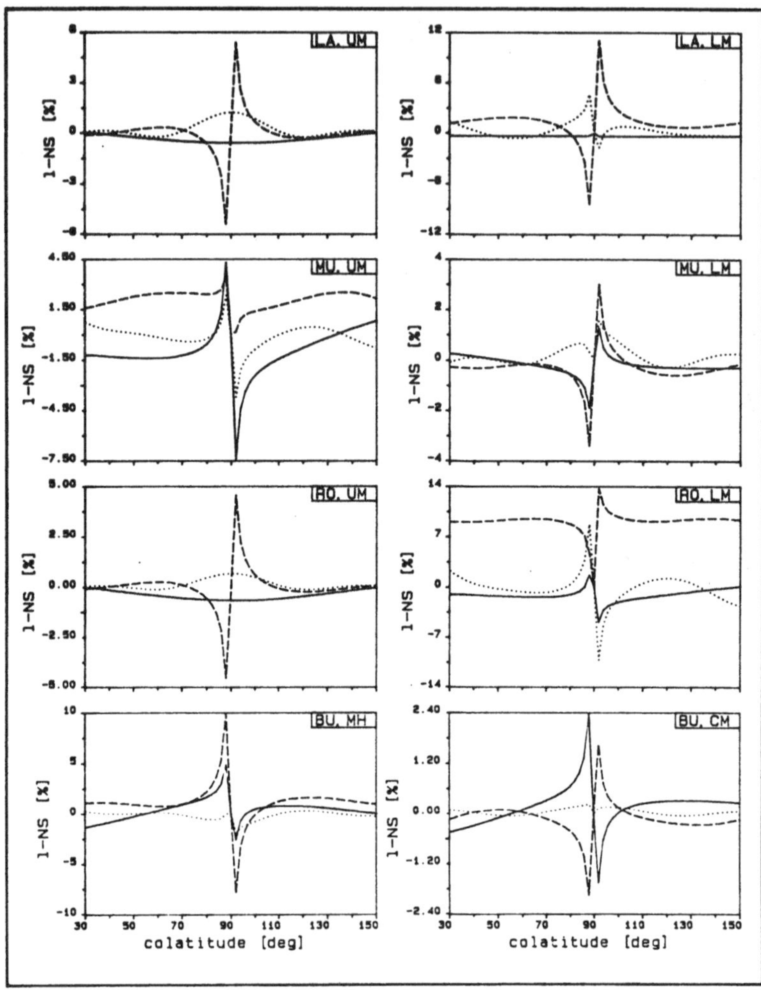


**Figure 4.** Same as Figure 1. but for the north-south component of Love number l.

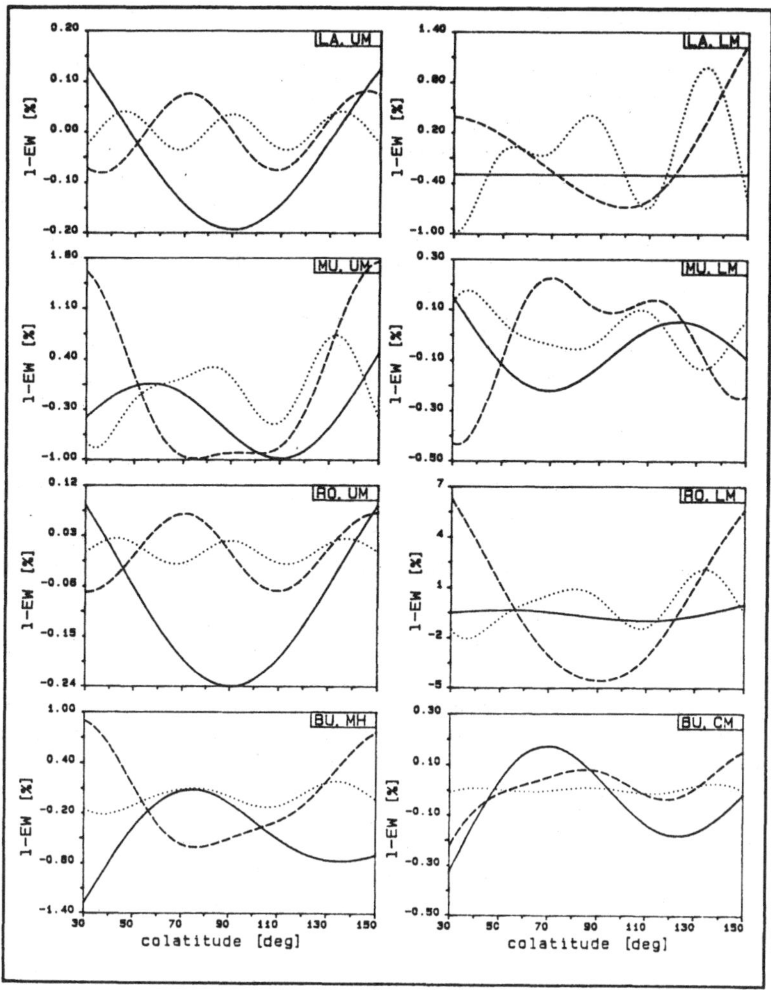


**Figure 5.** Same as Figure 1, but for the east-west component of Love number l.

### 7.8 M2 tidal residuals caused by large-scale mantle heterogeneities

As mentioned above, seismic tomography predicts large-scale mantle heterogeneities in the seismic velocities of the order of up to 2%, with an exception in the region of the asthenosphere, where the maximal anomalies can reach up to 8%. It has been shown that the mantle heterogeneities lead to various interesting effects in the tidal parameters. But the exact magnitudes of these effects are not yet known. In this section we will construct a three-dimensional model for mantle rheology based on the results of seismic tomography and calculate the induced global residuals in the semi-diurnal earth tides by using the perturbation method.

A major problem in making use of the contemporary three-dimensional model for the distribution of seismic velocities is the imperfect mapping of the complete rheological parameters because of the insufficient resolution. For our purpose we need the distribution of the three independent parameters, i.e. density $\rho$, bulk modulus k and shear modulus $\mu$. A possible way to solve this problem is to assume that the seismic velocities and the density vary systematically with each other. This assumption seems to be reasonable for heterogeneities due to temperature variations (Masters et al., 1982; Anderson et al., 1968). However, it has been shown that for a rigid earth model, using reasonable density anomalies assigned to the velocity anomalies, the calculated low degree geoid undulations are several times larger than the observed ones and may even have opposite signs (Dziewonski et al., 1977). Richards & Hager (1984) treat the density heterogeneities defined by tomography as internal loads. These internal loads induce a viscous circulation which deflects the existing density interfaces such as the core-mantle boundary and the outer surface. This dynamics of viscous flow in the mantle could reduce the long-wavelength equivalent geoid anomalies from those resulting from the driving density contrasts alone and could even lead to a reversal in sign. But the degree and characteristics of the reduction depend entirely on the model of the steady state viscosity in the mantle. Here we will neglect the deflection of the density interfaces induced by the dynamical viscous flow and assume the P wave velocity and the density to vary systematically with the S wave velocity according to the following empirical relationships:

$$(7\text{-}28) \qquad \delta V_p^2 / V_p^2 = 0.8\,\delta V_s^2 / V_s^2 , \qquad \delta\rho / \rho = 0.2\,\delta V_s^2 / V_s^2 .$$

(Masters et al., 1982; Anderson et al., 1968). Furthermore, in the mantle it holds approximately (considering PREM)

$$(7\text{-}29) \qquad V_p^2 / V_s^2 \sim 3.30 \quad \text{and} \quad \lambda / \mu \sim 1.33 .$$

The assumptions and approximations above make it possible to completely model lateral heterogeneities in all of the three above mentioned rheological parameters, if any one of the variations $\delta V_p$ or $\delta V_s$ is known. For example,

(7-30) $\delta\rho / \rho \sim 0.20\, \delta V_s^2 / V_s^2 \sim 0.25\, \delta V_p^2 / V_p^2$

(7-31) $\delta k / k \sim 0.87\, \delta V_s^2 / V_s^2 \sim 1.08\, \delta V_p^2 / V_p^2$

(7-32) $\delta\mu / \mu \sim 1.20\, \delta V_s^2 / V_s^2 \sim 1.50\, \delta V_p^2 / V_p^2$ .

These empirical relationships allow us to construct a three dimensional model for all three rheological parameters in the mantle by making use of the results of seismic tomography. Here we use model L02.56 of Dziewonski (1984) for the lateral heterogeneity in P velocity up to degree and order 6 and model M84C of Woodhouse & Dziewonski (1984) for the lateral heterogeneity in squared S velocity and the perturbation in crustal thickness (undulation of the Moho-discontinuity) up to degree and order 8. The effect of elliptical stratification and rotation is omitted here since it has been considered separately in Section 7.4.

For our three dimensional earth's model we calculate global residuals of different tidal observables in the semi-diurnal band. The residuals or total variations in the vertical surface displacement, for example, have been defined in (6-6) and are described by the cosine and sine part, which correspond to the real and imaginary part of the variations in the complex representation. The results show that the tidal residuals caused by these large-scale lateral heterogeneitis are very small. In Table 7 we find (in the columns containing the result calculated without amplification effects) the M2-residuals of the surface displacement to be on the order of 0.5 mm, the Eulerian potential about 10 gal·cm, the gravity perturbation about 30 ngal and the perturbation in tilt about 0.05 nradian. Figure 6 shows the percentage effect of different tidal parameters at the 40°-latitude profile. For both, the north-south and the east-west components of the Love number l, variations are of the order of one per cent in the amplitude and one degree in the phase shift. The effects in the Love numbers h and k and also in the two tilt factors are smaller by one order of magnitude than those in the Love number l. The variation in the gravimetric factor at this profile is not significant.

It is obvious that these small effects cannot exceed the error level of the contemporary techniques of measurement. Therefore, the influences of the large-scale lateral heterogeneities in the mantle on earth tides can be neglected, if the heterogeneities predicted by seismic tomography are also valid in the tidal band.

However, if the lateral heterogeneities are induced partly by differences of the ambient condition, which are essentially differences of temperature, they could be considerably stronger at the tidal periods than at seismic periods (Wang & Zschau, 1986). Assuming that a small anomaly $\delta\mu$ in the shear modulus induced by a perturbation of Zschau's parameter $\delta z$, is known for a given reference seismic frequency $\omega_o$, to first order we can express

$$(7\text{-}33) \qquad \delta\mu(z, \omega_o) = \delta z \cdot \frac{\partial}{\partial z}\mu(z, \omega_o).$$

At another frequency $\omega$, this anomaly will disperse to

$$(7\text{-}34) \qquad \delta\mu(z, \omega) = \delta z \cdot \frac{\partial}{\partial z}\mu(z, \omega)$$

$$= \delta\mu(z, \omega_o) \cdot \frac{\frac{\partial}{\partial z}\mu(z, \omega)}{\frac{\partial}{\partial z}\mu(z, \omega_o)} .$$

By using (7-26), we find

$$(7\text{-}35) \qquad \delta\mu(z, \omega) = \delta\mu(z, \omega_o)\left\{ \frac{F(z, \omega)}{F(z, \omega_o)} + \frac{\frac{\partial}{\partial z}\big(F(z, \omega)/F(z, \omega_o)\big)}{\frac{\partial}{\partial z}\ln\big(\mu_o(z)\big) + \frac{\partial}{\partial z}\ln\big(F(z, \omega_o)\big)} \right\}.$$

Supposing that

$$(7\text{-}36) \qquad \frac{\partial}{\partial z}\ln\big(\mu_o(z)\big) = \beta\frac{\partial}{\partial z}\ln\big(F(z, \omega_o)\big)$$

relation (7-35) is reduced to

$$(7\text{-}37) \qquad \delta\mu(z, \omega) = \delta\mu(z, \omega_o)\cdot G(z; \omega, \omega_o).$$

Here G is called the amplification factor and defined by

$$(7\text{-}38) \qquad G(z; \omega, \omega_o) = \frac{\beta}{1+\beta}\,\frac{F(z, \omega)}{F(z, \omega_o)} + \frac{1}{1+\beta}\,\frac{\frac{\partial}{\partial z}F(z, \omega)}{\frac{\partial}{\partial z}F(z, \omega_o)} .$$

The parameter $\beta$ is not only dependent on composition and phase of the solid material, the ambient temperature and pressure but also on the rheological model to be used. In fact, no rheological model can describe the visco-elastic behaviour of solids at all frequencies. The unrelaxed modulus, as mentioned above, is in practice the high frequency modulus. In the following we assume that lateral mantle inhomogeneities essentially exhibit the visco-elastic dispersion modified by the inhomogeneous ambient condition, that is to say no significant influence of the ambient condition on the unrelaxed modolus ($|\beta| \ll 1$). Based on Zschau's relaxation theory, we find a significant amplification factor of the order of 3-5 and 5-10 for the semi-diurnal tides, supposing the seismic reference period to be 200 s and 30 s, respectively (Table 8). This amplification effect should be considered for extending the 3-D seismic earth model to the tidal band.

**Table 7.** Calculated extrema of the global M2 tidal residuals caused by the large-scale lateral mantle heterogeneities, the Moho- and the geoid undulations. The effects increase by several times if the temperature-induced amplification of the heterogeneity in the shear modulus is considered in the tidal period by assuming the seismic reference period to be 200s.

| tidal observables | with amplification effect | | | | without amplification effect | | | |
|---|---|---|---|---|---|---|---|---|
| | cosine part | | sine part | | cosine part | | sine part | |
| | max. | min. | max. | min. | max. | min. | max. | min. |
| $\Delta u^{(n)}$ (mm) | 0.62 | -0.62 | 0.59 | -0.51 | 0.14 | -0.19 | 0.16 | -0.11 |
| $\Delta u^{(s)}$ (mm) | 1.59 | -1.46 | 1.60 | -1.61 | 0.40 | -0.31 | 0.41 | -0.49 |
| $\Delta u^{(e)}$ (mm) | 1.31 | -1.57 | 1.69 | -1.88 | 0.45 | -0.50 | 0.49 | -0.49 |
| $\delta\varphi$ (gal·cm) | 27.1 | -21.0 | 19.5 | -20.1 | 11.3 | -11.6 | 8.8 | -13.2 |
| $\Delta g_t^{(n)}$ (ngal) | 108. | -173. | 123. | -104. | 30. | -24. | 28. | -17. |
| $\Delta t_t^{(s)}$ (nrad.) | 0.19 | -0.21 | 0.20 | -0.21 | 0.04 | -0.05 | 0.04 | -0.05 |
| $\Delta t_t^{(e)}$ (nrad.) | 0.40 | -0.34 | 0.28 | -0.35 | 0.05 | -0.05 | 0.06 | -0.06 |

Table 7 shows that the tidal residuals increase significantly by introducing this amplification effect in the shear modulus anomalies. By assuming the central seismic reference period to be 200 s and the shear modulus anomalies to be fully induced by temperature differences, residuals of the M2-tide could be intensified by a factor of 5-7 in the vertical displacement, in the gravity and in the east-west component of tilt, by a factor of 3-4 in the north-south component of tilt and in both components of the horizontal displacement, and by a factor of about 2 in the Eulerian potential. From this feature and from the results of the sensitivity studies in Section 7.7 it is to be concluded that the influences of lateral heterogeneities on earth tides are essentially traceable to those of the shear modulus in the upper mantle.

**Table 8.** Amplification of the shear modulus anomaly at the tidal period (M2) with respect to that at seismic periods of 30s and 200s. The determination of the amplification factor is based on the assumption that $\beta = 0$ in (7-38).

| depth | $\rho_o T / \Phi$ | amplification factor G (real, imaginary) | | | |
|---|---|---|---|---|---|
| (km) | ($10^{-8}$cgs) | M2 / 30s | | M2 / 200s | |
| 24- 60 | 0.80 | 4.5892 | -1.4944 | 3.0582 | -0.9958 |
| 60- 80 | 1.08 | 6.5978 | -2.7066 | 3.9925 | -1.6378 |
| 80- 150 | 1.25 | 8.0916 | -3.7149 | 4.6366 | -2.1287 |
| 150- 220 | 1.37 | 9.2470 | -4.4204 | 5.1261 | -2.4505 |
| 220- 310 | 1.20 | 7.6219 | -3.3785 | 4.4376 | -1.9671 |
| 310- 400 | 1.16 | 7.2589 | -3.1197 | 4.2803 | -1.8396 |
| 400- 500 | 1.15 | 7.1235 | -3.0623 | 4.2148 | -1.8119 |
| 500- 670 | 1.14 | 7.0992 | -3.0321 | 4.2107 | -1.7984 |
| 670- 771 | 1.21 | 7.7126 | -3.4395 | 4.4756 | -1.9959 |
| 771-1271 | 1.16 | 7.2589 | -3.1197 | 4.2803 | -1.8396 |
| 1271-1771 | 1.05 | 6.3621 | -2.6004 | 3.8850 | -1.5879 |
| 1771-2271 | 1.00 | 5.9689 | -2.3078 | 3.7103 | -1.4344 |
| 2271-2741 | 0.96 | 5.6711 | -2.1714 | 3.5720 | -1.3677 |
| 2741-2891 | 0.99 | 5.8945 | -2.2612 | 3.6744 | -1.4096 |

For the semi-diurnal tide M2 at the 40°-latitude profile, variations in the horizontal surface displacement are about 1% in amplitude and 1° in the phase shift (Fig. 6). Including the amplification effect the variations reach up to 4-8% in amplitude and 2-4° in phase shift (Fig. 7). Variations in the other tidal observables do not exceed the 0.2% level without the amplification effect, but they reach up to 0.5% by including this effect.

The global descriptions for all calculated tidal residuals are shown in Figures 8-14. From these figures, we see that the 40°-latitude profile, around which most earth tide measurements take place, is not favourable for a conclusive judgement of the large-scale mantle heterogeneities from tidal gravity, tilt and potential data. Unfortunately, the relatively large residuals of these tidal observables are in the oceanic region. Similar characteristics of distribution can be found for the residuals in the vertical displacement and in the Eulerian potential and for the gravity and the east-west component of tilt. The former are dominated by the zonal components of spherical harmonics. Residuals in both the north-south and the east-west component of the horizontal displacement are more significant than those in the vertical displacement, and they are dominated by the sectorial harmonics. Because the radial distribution of the lateral heterogeneities varies in the mantle, it is impossible to find simple correlations between the tidal perturbations and the mantle heterogeneities.

It should be pointed out that the tidal residuals due to the large-scale mantle heterogeneities given here are only a rough estimate. More reliable estimations of the effects rely on the sufficiently perfect mapping of the mantle by seismic tomography and the knowledge of mantle inelasticity, especially of the temperature induced relaxation and dispersion. In the model calculations above, the seismic reference period has been chosen to be 200s for determining the visco-elastic amplification factor. It is about the central period of the free oscilation data which were applied to construct the reference earth model PREM. The periods of body waves in seismic tomography are much shorter. Consequently, the visco-elastic amplification of lateral heterogeneities in the tidal band could be considerably stronger than the factor of 3-5. Also the deflection of the interior density interfaces induced by the possible viscous flow could contribute additional effects on earth tides.

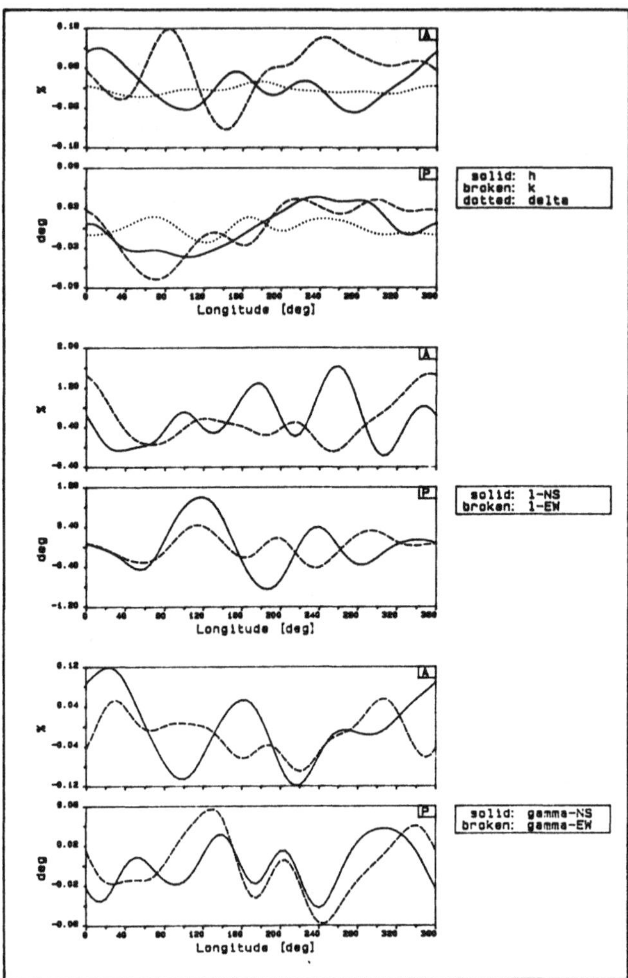


**Figure 6.** Influences of large-scale mantle heterogeneities (without the visco-elastic amplification) on semi-diurnal earth tides at the 40°N-latitude profile. The influences are given for the amplitude A (in percent) and for the phase B (in degree) of different tidal parameters. The undisturbed tidal parameters are determined for the spherically symmetric starting earth model PREM.

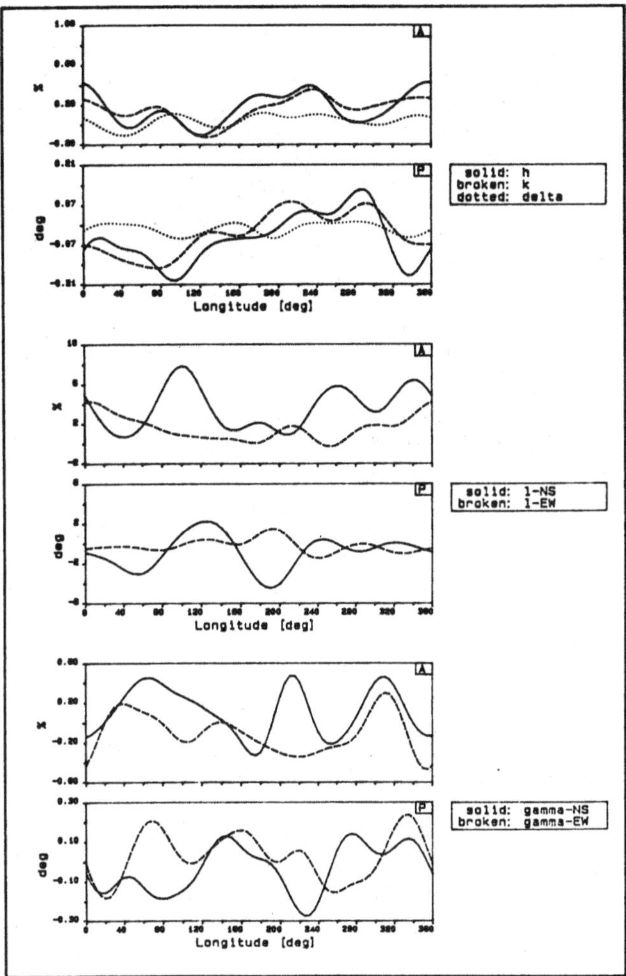


**Figure 7.** Same as Figure 6, but with the visco-elastic amplification in the lateral heterogeneities of the shear modulus taken into account.

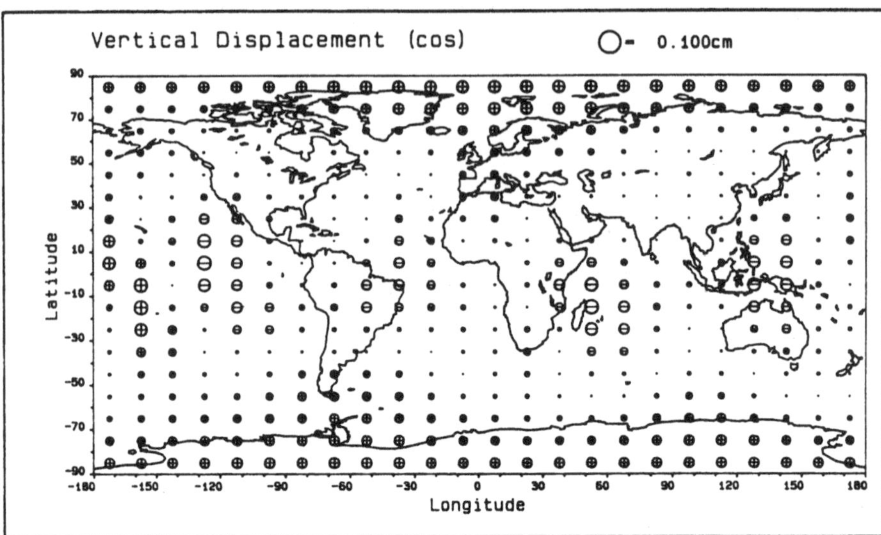


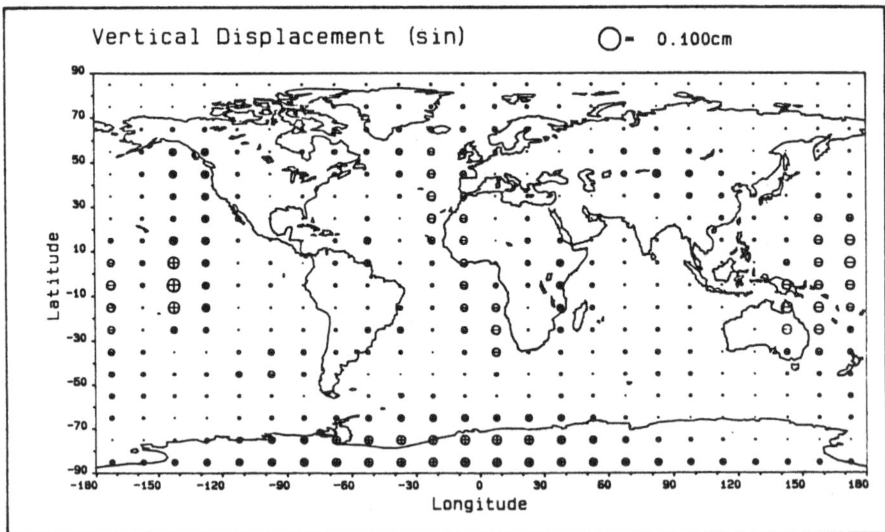


**Figure 8.** Global M2-residuals of the vertical surface displacement caused by large-scale mantle heterogeneities as well as Moho- and geoid undulations. In order to extend the three-dimensional mantle model predicted by seismic tomography to the semi-diurnal tidal period, the visco-elastic amplification in the heterogeneities of the shear modulus has been considered based on Zschau's theory of temperature induced visco-elastic relaxation.

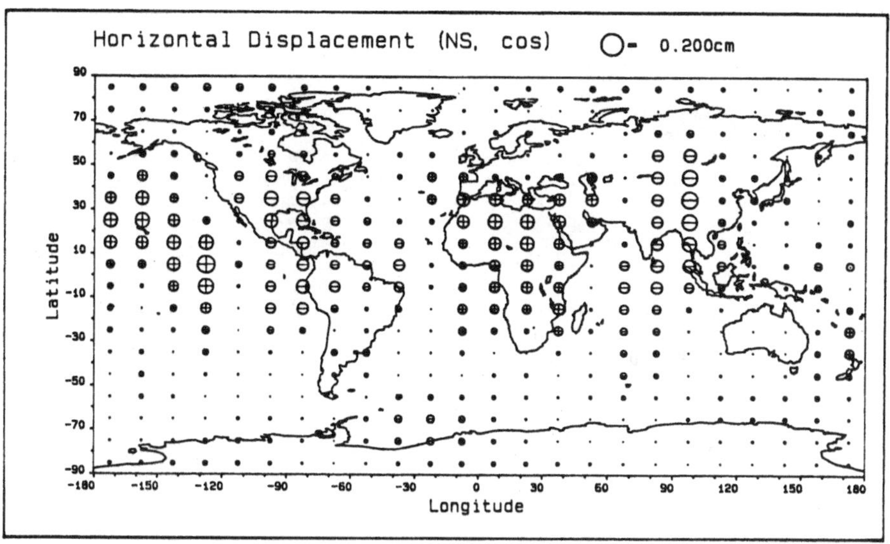


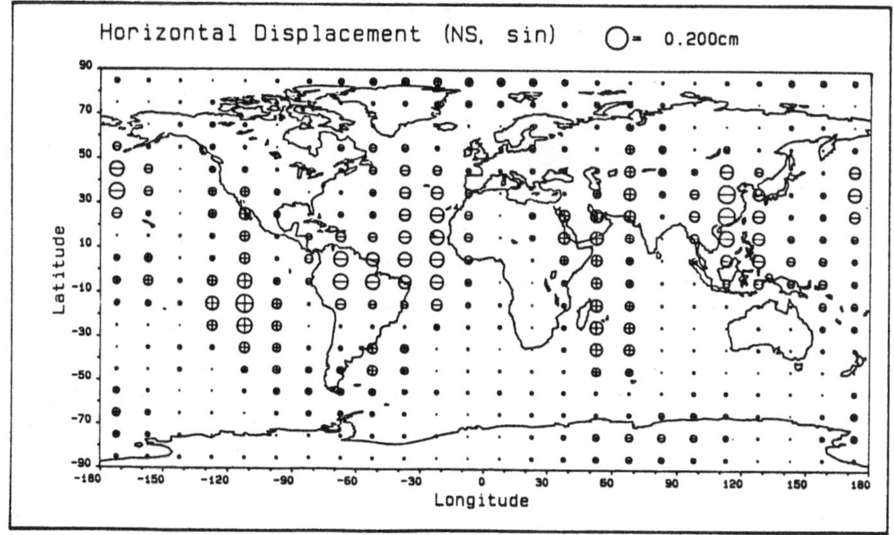


**Figure 9.** Same as Figure 8, but for the horizontal surface displacement in north-south direction (southward positive).

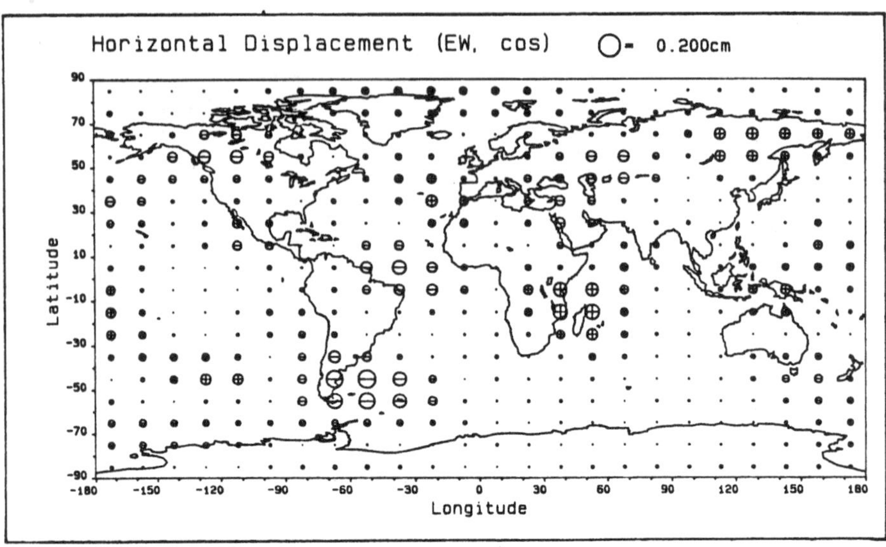


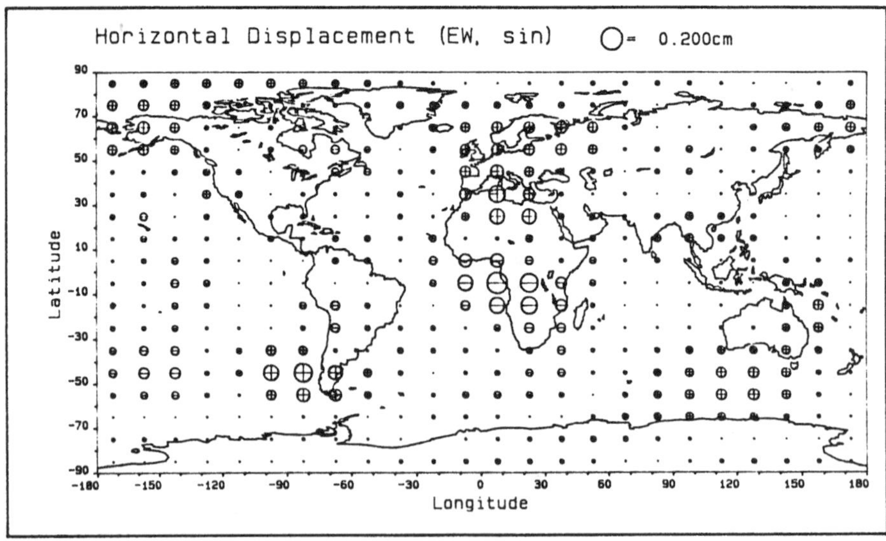


**Figure 10.** Same as Figure 8, but for the horizontal surface displacement in east-west direction (eastward positive).

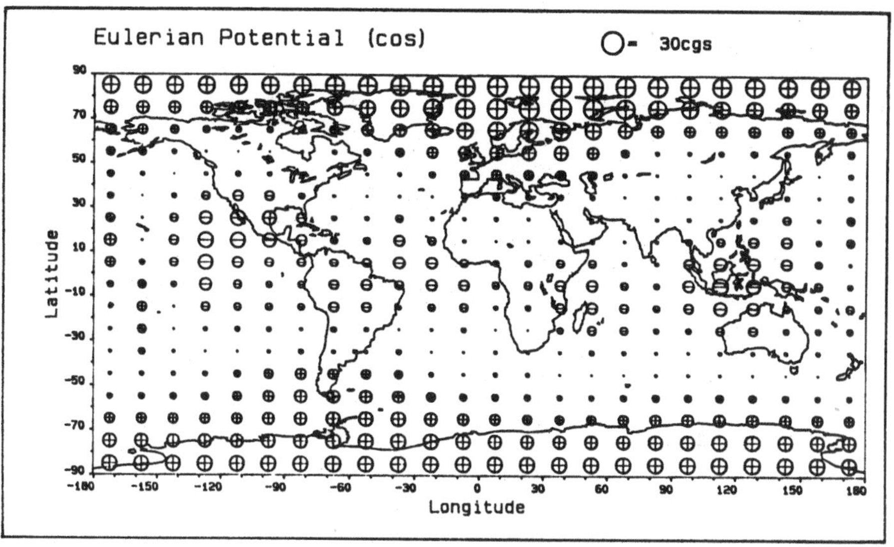


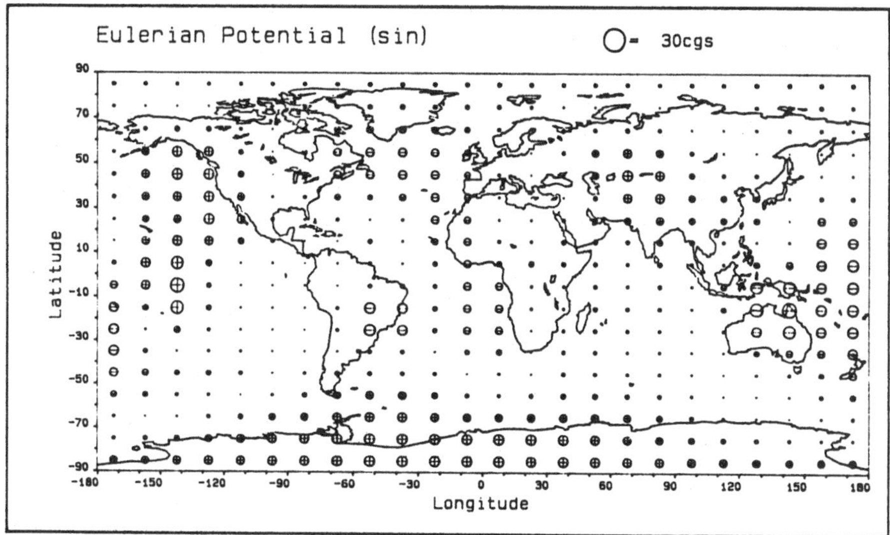


**Figure 11.** Same as Figure 8, but for the free space Eulerian potential related to the undisturbed spherical earth's surface.

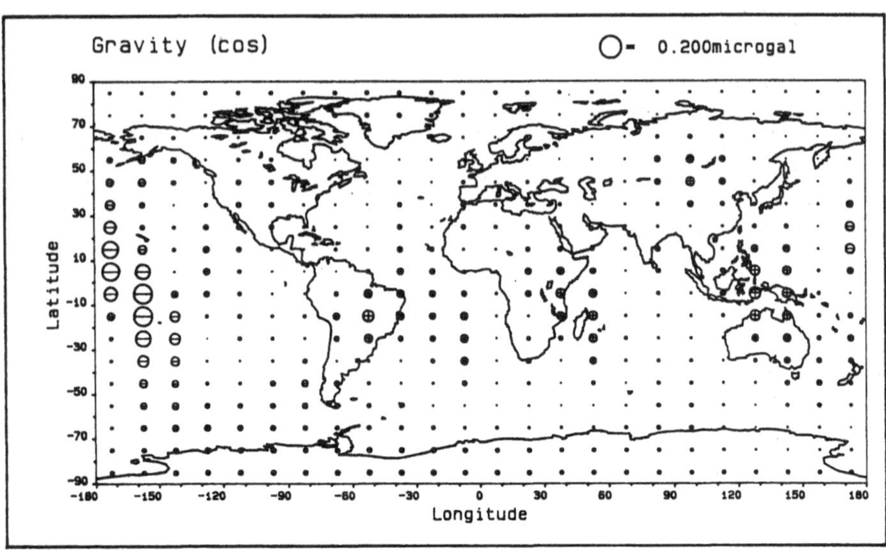


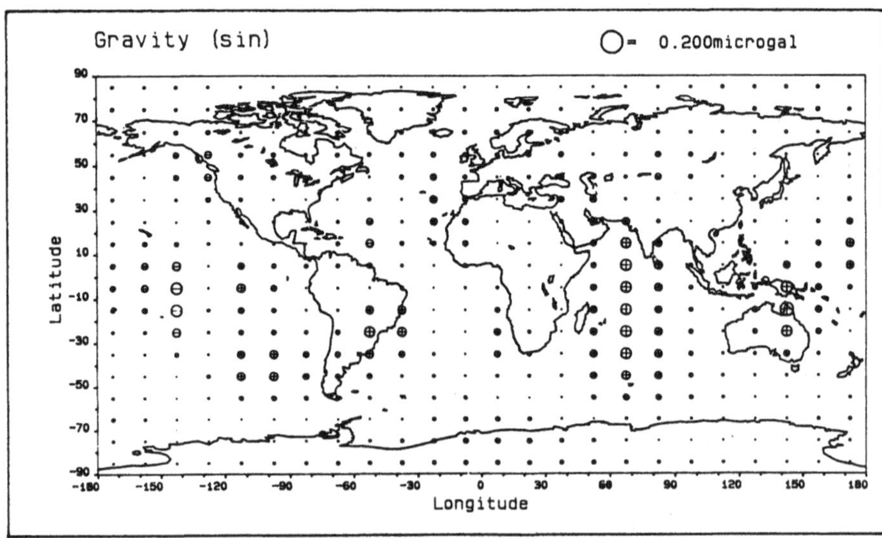


**Figure 12.** Same as Figure 8, but for tidal gravity.

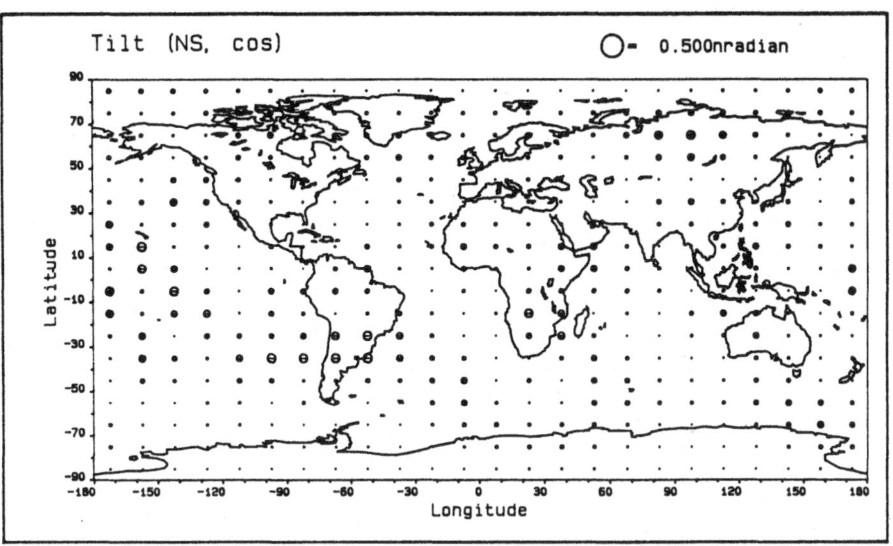


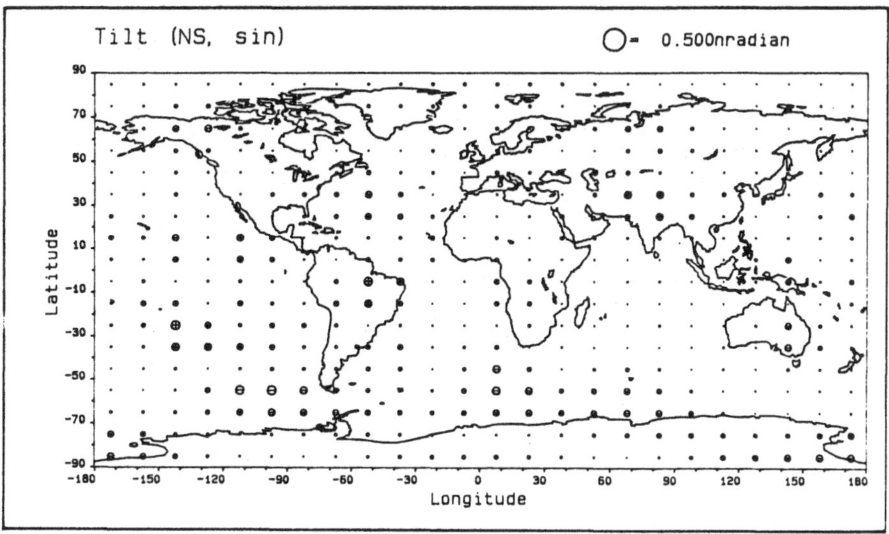


**Figure 13.** Same as Figure 8, but for tilt in north-south direction (southward positive).

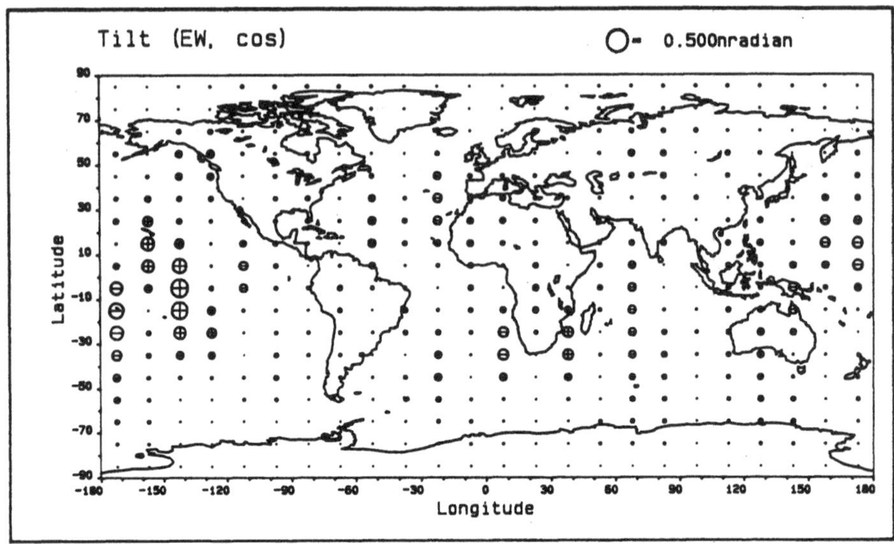


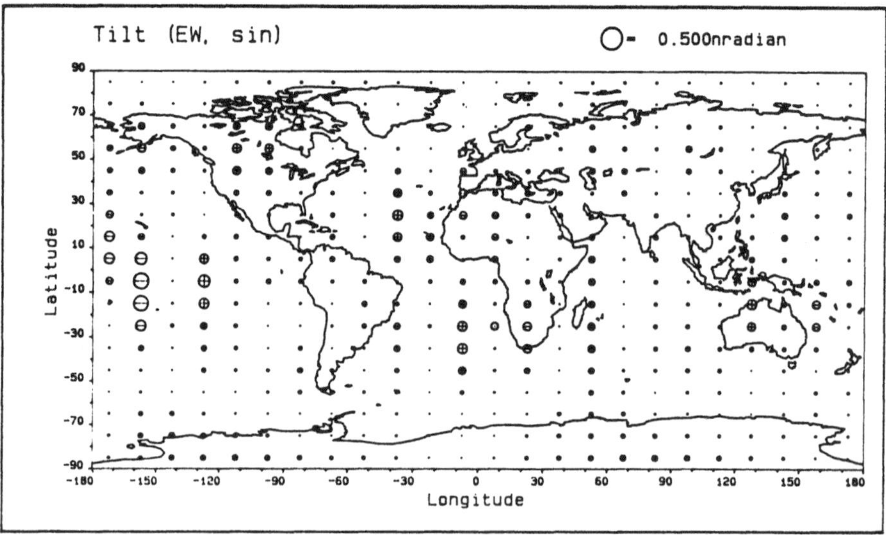


**Figure 14.** Same as Figure 8, but for tilt in east-west direction (eastward positive).

## Chapter 8. Summary

Earth tides are influenced by the earth's rotation as well as by elliptical stratification, inelasticity and lateral heterogeneities in the mantle. Tidal perturbations due to these small deviations from a spherically symmetric, non-rotating and elastic model earth can be modelled by the perturbation method. For perturbations of long wavelength, it is convenient to make use of expansion series of spherical harmonics for both the tidal observables and the model parameters. A principle advantage of the perturbation method applied here is that the expansion coefficients of the solution can be evaluated independently from each other by a semi-analytical procedure, while with a direct method, a large coupled system of differential equations would have to be solved.

To verify the validity of the computational procedure, the numerical results for several special examples have been compared with the corresponding analytical ones. From this we find that the numerical errors in determining the tidal perturbations due to the rotation and elliptical stratification of a homogeneous or fluid earth do not exceed one per mill.

With respect to body tides on an elliptical, rotating, and elastic earth, there are some significant discrepancies between our results and those given by Wahr (1981). The latitude dependences of the equivalent Love number k and of the gravimeter factor δ are only half as large as those in Wahr's results. If we use the modified definition of k and δ suggested by Dehant & Ducarme (1987), in which the theoretical tidal quantities are related to the elliptical surface instead of the spherical one, there is no significant latitude dependence of these two tidal parameters any more. However, the gravimeter factor shows an increase of 0.2%, 0.4% and 0.3% compared to the average δ-values given by Dehant & Zschau (1989) for Mf, $O_1$ and $M_2$, respectively.

The effect of inelasticity in the earth's mantle has been determined by both the perturbation method and the direct integration method. Based on Zschau's rheological model, mantle inelasticity in the main tidal band (semi-diurnal and diurnal periods) produces an increase of about 1% in Love numbers h and k and more than 2% in Love number l, and a small phase shift of 0.2° for h and k and 0.5° for l. At longer periods the dispersion of the Love numbers becomes more significant. For the Chandler wobble (period of 435 days) the increase of the Love numbers is 4-5% in h and k and more than 10% in l, and the phase shift reaches up to 1° for h and k and 5° for l, respectively.

In a lateral heterogeneous earth the equivalent tidal parameters become complex even in the purely elastic case. This is due to axial asymmetry which introduces phase shifts in the tidal observables. Both the amplitude and the phase shift of the equivalent tidal parameters do not only depend on the latitude but also on the longitude, and Love number l and the tilt factor γ are anisotropic. The nodal lines of the tidal response of the earth do not coincide with those of the corresponding theoretical tides any more, and consequently, the tidal parameters as defined traditionally can become infinite.

The perturbations in the Love numbers h and k are relatively sensitive to large-scale lateral heterogeneities in the shear modulus of the upper mantle. To resolve this heterogeneity on the level of about one per cent, an accuracy of at least a few tenths of a per cent is required in the global determination of the Love numbers h and k. The effect on the gravimeter factor δ is systematically smaller. The reason for this is the similar behavior of the Love numbers h and k which decreases the total effect on δ to a large extent because the Love numbers enter in the gravimeter factor with opposite sign. It is concluded that mantle heterogeneities, as estimated by the seismic tomography, will not be seen in the tidal gravity data even if the accuracy of the global determination of the gravimeter factor is improved by one order in the future. A large percentage effect is to be expected in Love number l. It is in particular sensitive to mantle heterogeneities of short wavelengths. This suggests that regional or local heterogeneities could cause significant effects in the horizontal surface displacement.

Seismic tomography predicts the seismic large-scale lateral heterogeneities in the earth's mantle to be smaller than 2%, with the exception of the region of the asthenosphere where the maximum anomalies in the seismic velocities can reach up to 8%. From our model calculations the corresponding effects on earth tides seem to be very small. However, we suggest that the lateral heterogeneities might show dispersion if they are partly induced by temperature differences. This means, anomalies in the shear modulus due to temperature differences are more significant at lower frequencies than at higher frequencies. Consequently, the lateral heterogeneities could be considerably stronger at tidal periods than at seismic periods. Thus, in determining tidal perturbations from seismic models, the visco-elastic amplification in the temperature-induced heterogeneities of the shear modulus should be considered. The amplification factor is, in general, a complex value. In the semi-diurnal band, its amplitude is about 3-5 and 5-10 by assuming the seismic reference period of the PREM to be 200 s and 30 s, respectively.

By introducing temperature-induced visco-elastic dispersion in the seismic lateral heterogeneities, the influences on e.g. the semi-diurnal tides increase significantly. By assuming the seismic reference period to be 200 s and the shear modulus heterogeneites to be fully induced by temperature differences, tidal residuals in the vertical displacement and in gravity are intensified by a factor of 5-7 and reach up to the level of 1 mm and 200 ngal, respectively. In the horizontal displacement the residuals are increased by a factor of 3-4 and to the level of 2 mm, and in the Eulerian potential they are increased by a factor of about 2 and to the level of 30 gal·cm, respectively. At the 40°-latitude profile, perturbations in Love number l number h and k and in the tilt factor are about 0.5% in amplitude and 0.3° in phase shift. In the gravimeter factor there are only small perturbations of 0.1% in amplitude and 0.05° in phase.

It should be noted that this order of magnitude is the effect on the tidal component $M_2$ alone, and is solely due to very large-scale lateral heterogeneities (up to degree and order 8), which correspond to wavelengths of several thousand kilometers. Furthermore, the heterogeneities of density, the bulk modulus and the shear modulus have been only very roughly estimated. The effect due to the deflection of the interior boundaries of discontinuity induced by the possible viscous flow has been neglected. In the model calculations, the amplification factor has been taken to be 3-5, deduced from Zschau's rheological model for temperature induced relaxation by assuming the seismic reference period of 200s. This is a conservative estimate. By taking the reference periods to be the much shorter periods of the body waves applied to seismic tomography, this visco-elastic amplification is considerably stronger. Thus, if the rheological parameters of the earth's mantle can be much more sufficiently modelled, the total effect on the complete earth tides may be much larger.

The influences of the earth's rotation as well as its elliptical stratification, inelasticity and heterogeneities on earth tides are just starting to be considered. The experimental verification of the theoretical models requires a high accuracy of global earth tide measurements, at least a few tenths of a per cent. This accuracy can, in principle, be achieved with terrestrial instruments if superconducting gravimeters are used, but such measurements provide only information on the gravimeter factor. In this quantity, as pointed out above, the Love numbers h and k unfortunately enter with opposite sign, and as consequence several interesting influences of mantle anelasticity and heterogeneities cancel to a large extent. The separate determination of the Love numbers can be achieved by modern space techniques. In fact, recent observations from Very Long Baseline Interferometry (VLBI) have estimated the station dependence of Love numbers to be on the level of 1% for h and 25% for l (Carter et al., 1985; Ryan et al., 1986). This could already reflect lateral heterogeneities in the mantle. Also Satellite Laser Ranging (SLR) and Lunar Laser Ranging (LLR) have the same capability of providing information on station dependent Love numbers. However, a conclusive judgement requires not only an improvement of the accuracy of observations but also a better global coverage with VLBI as well as Laser Ranging stations.

## Appendix A. Theoretical Background for Tidal Deformations of a Self-gravitating, Spherically Symmetric and Elastic Earth

The linearized differential equations of infinitesimal dynamical deformations in a self-gravitating, spherically symmetric and elastic earth consist of equation of momentum conservation and Poisson's equation (Backus, 1967):

$$\text{(A-1)} \quad \begin{cases} \rho \ddot{\mathbf{u}} = \rho\{ \nabla(\varphi + u_r \cdot dV/dr) - (\nabla\cdot\mathbf{u}) dV/dr\,\mathbf{e}_r \} + \nabla\cdot\boldsymbol{\sigma} \\ \nabla^2 \varphi = 4\pi G \nabla\cdot(\rho\,\mathbf{u}) \end{cases}$$

where $\rho$ is density, V the gravity potential, $\mathbf{u}$ the infinitesimal displacement vector, $\varphi$ the infinitesimal potential due to exterior tide-generating forces and interior density redistribution during deformations, and $\boldsymbol{\sigma}$ the infinitesimal stress tensor, expressed by the relation

$$\text{(A-2)} \quad \boldsymbol{\sigma} = \lambda(\nabla\cdot\mathbf{u})\mathbf{I} + \mu[\nabla\mathbf{u} + (\nabla\mathbf{u})^T] .$$

Equations (A-1) are to be solved under the interior boundary conditions

$$\text{(A-3)} \quad [\mathbf{e}_r \cdot \boldsymbol{\sigma}]_-^+ = 0;$$

$$\text{(A-4)} \quad \begin{cases} [\mathbf{u}]_-^+ = 0, & \text{if welded;} \\ [\mathbf{n}\cdot\mathbf{u}]_-^+ = 0, & \text{if frictionless;} \end{cases}$$

$$\text{(A-5)} \quad [\partial\varphi/\partial r - 4\pi G \rho u_r]_-^+ = 0.$$

and the surface conditions

$$\text{(A-6)} \quad \mathbf{e}_r \cdot \boldsymbol{\sigma} = \mathbf{b}$$

$$\text{(A-7)} \quad [\partial\varphi/\partial r - 4\pi G \rho u_r]^- = [\partial\varphi/\partial r]^+$$

In addition, all quantities must be regular at the origin ($r = 0$).

In general, the surface conditions may be expanded in terms of the so called spheroidal and toroidal modes, which can be expressed further in terms of spherical harmonics or their surface gradients. From free oscillations we know that for a spherically symmetric earth the spheroidal and toroidal solutions are decoupled. The main results given here are adopted from Wiggins (1968). However, some special treatments of the equations in the liquid core, the boundary conditions at frictionless boundaries in the case of very low frequency and the starting vectors near the earth's origin are new.

A.1 Toroidal solutions

The toroidal solutions are characterized by the vanishing of the dilatation $\nabla \cdot \mathbf{u}$ and the radial component of displacement $u_r$. With the interior density remaining undisturbed and with no toroidal part of the gravitational potential from exterior tide-generating bodies, there is no infinitesimal potential $\varphi$. The solution for degree n and order m of spherical harmonics is commonly given in the form

(A-8) $$\mathbf{u} = W_n(r)\,\mathbf{e}_r \times \nabla_1 Y_{nm}(\vartheta, \lambda)\, e^{i\omega t}.$$

Substituting this solution into the equations of motion we obtain an ordinary differential equation of second order for the radially dependent coefficient $W_n$. For numerical convenience this second order equation will be transformed to a set of two simultaneous equations of first order. Defining transformations

(A-9) $$\mathbf{y} = \begin{bmatrix} y_1 \\ y_2 \end{bmatrix} = \begin{bmatrix} W_n \\ \mu(\dot{W}_n - W_n/r) \end{bmatrix}$$

where $y_2$ is the radial coefficient of the shear stresses $\sigma_{r\vartheta}$ and $\sigma_{r\varphi}$, the first order equation system can be written as

(A-10) $$d\mathbf{y}/dr = \mathbf{A}\mathbf{y}$$

where $\mathbf{A}$ is a $2 \times 2$-matrix with coefficients

$$A_{11} = 1/r$$

$$A_{12} = 1/\mu$$

$$A_{21} = (n-1)(n+2)\mu/r^2 - \rho\omega^2$$

$$A_{22} = -3/r .$$

However, the equation system (A-9) is poorly conditioned for numerical computation. This difficulty can be surmounted by the following transformations of the dependent variables:

$$\text{(A-11)} \quad \tilde{\mathbf{y}} = r^{1/2} \begin{bmatrix} y_1 \\ r y_2/(n+1) \end{bmatrix}.$$

The transformed system then is

$$\text{(A-12)} \quad d\tilde{\mathbf{y}}/dz = \tilde{\mathbf{A}}\,\tilde{\mathbf{y}}$$

where the matrix $\tilde{\mathbf{A}}$ has coefficients

$$\tilde{A}_{11} = 3/[2(n+1)]$$

$$\tilde{A}_{12} = 1/\mu$$

$$\tilde{A}_{21} = [(n-1)(n+2)\mu - \rho\omega^2 r^2]/(n+1)^2$$

$$\tilde{A}_{22} = -\tilde{A}_{11}$$

and the former independent variable r is related to z by the relation

$$r = \exp[z/(n+1)].$$

The boundary conditions are that the variable vector $\mathbf{y}$ or $\tilde{\mathbf{y}}$ must be continuous through all welded boundaries. Since there is no shear stress in the liquid medium and the shear stress vanishes at frictionless boundaries, the toroidal deformations will occur only in the solid layers underneath the surface. The integration, therefore, can be started at the core-mantle boundary with the starting vector

$$\text{(A-13)} \quad \tilde{\mathbf{y}}_o = \begin{bmatrix} C \\ 0 \end{bmatrix}$$

where C is a non-zero constant. In the case of free oscillations this constant can be chosen arbitrarily. One finds the normal modes by varying the frequency $\omega$ until the homogeous stress condition, $y_2 = 0$, at the earth's surface is satisfied. The corresponding frequencies are the eigenfrequencies. For our purpose, however, the forced tidal deformations are considered and the frequency $\omega$ is, in general, much lower than the eigenfrequencies. In this case, this constant should be determined by the inhomogeous surface stress conditions. Because the system is linear, we may let, at first, C = 1 and integrate the equations from the core-mantle boundary to the earth's surface, then determine the factor, required to match the boundary condition at the surface.

A.2 Spheroidal solutions

The spheroidal solutions of degree n and order m are expressed as

(A-14)
$$\begin{cases} \mathbf{u} = \{ H_n(r)\mathbf{e}_r + T_n(r)\nabla_1 \} Y_{nm}(\vartheta, \lambda)e^{i\omega t} \\ \varphi = R_n(r)Y_{nm}(\vartheta, \lambda)e^{i\omega t}. \end{cases}$$

The second order differential equations of the dependent variables $H_n$, $T_n$ and $R_n$ can be transformed into a set of six simultaneous equations of first order by the substitutions

(A-15)
$$\mathbf{y} = \begin{bmatrix} y_1 \\ y_2 \\ y_3 \\ y_4 \\ y_5 \\ y_6 \end{bmatrix} = \begin{bmatrix} H_n \\ \lambda X_n + 2\mu \dot{H}_n \\ T_n \\ \mu(\dot{T}_n - T_n/r + H_n/r) \\ R_n \\ \dot{R}_n - 4\pi G\rho H_n \end{bmatrix}$$

where

(A-16)
$$X_n = \dot{H}_n + 2H_n/r - n(n+1)T_n/r$$

is the radial coefficient of the dilatation $\nabla \cdot \mathbf{u}$. The transformed equation system can be written in the form

(A-17)
$$d\mathbf{y}/dr = \mathbf{A}\mathbf{y}$$

where **A** is a 6×6-matrix with the coefficients

$$A_{11} = -2\lambda/[r(\lambda + 2\mu)]$$

$$A_{12} = 1/(\lambda + 2\mu)$$

$$A_{13} = n(n+1)\lambda/[r(\lambda + 2\mu)]$$

$$A_{21} = 4\mu(3\lambda + 2\mu)/[r^2(\lambda + 2\mu)] - 4\rho g/r - \rho\omega^2$$

$$A_{22} = -4\mu/[r(\lambda + 2\mu)]$$

$$A_{23} = n(n+1)\{\rho g/r - 2\mu(3\lambda + 2\mu)/[r^2(\lambda + 2\mu)]\}$$

$$A_{24} = n(n+1)/r$$

$$A_{26} = -\rho$$

$$A_{31} = -1/r$$

$$A_{33} = 1/r$$

$$A_{34} = 1/\mu$$

$$A_{41} = A_{23}/[n(n+1)]$$

$$A_{42} = -\lambda/[r(\lambda + 2\mu)]$$

$$A_{43} = 2\mu[2n(n+1)(\lambda + \mu)/(\lambda + 2\mu) - 1]/r^2 - \rho\omega^2$$

$$A_{44} = -3/r$$

$$A_{45} = -\rho/r$$

$$A_{51} = 4\pi G\rho$$

$$A_{56} = 1$$

$$A_{63} = -4\pi G\rho n(n+1)/r$$

$$A_{65} = n(n+1)/r^2$$

$$A_{66} = -2/r$$

all other $A_{ij} = 0$.

In the case of n = 0, there is no perturbation of the gravitational field in the earth's exterior because of conservation of the total mass. In the earth's interior, $y_6$ = 0. And the variables $y_3$ and $y_4$ have no meaning. The equations reduce to

$$\text{(A-18)} \quad \left| \begin{array}{l} dy_1/dr = A_{11} y_1 + A_{12} y_2 \\ dy_2/dr = A_{21} y_1 + A_{22} y_2 \\ dy_5/dr = A_{51} y_1 . \end{array} \right.$$

Note that the first two equations form a closed system. Because $y_5$ vanishes at the deformed earth's surface, we need to integrate the third equation only when the solutions in the earth's interior are of interest.

For n » 0, the shear stresses vanish in a liquid layer ($\mu$ = 0), i.e. $y_4$ = 0, and the equations concerning $dy_3/dr$ and $dy_4/dr$ will be replaced by the relation

$$\text{(A-19)} \quad y_2 + \rho(y_5 - gy_1 + r\omega^2 y_3) = 0 .$$

In the case of free oscillations this relation will be used as a solution of the equation concerning the variable $y_3$ and the equation system (A-17) reduces to a set of four simultaneous equations concerning the variables $y_1$, $y_2$, $y_5$ and $y_6$. However, this treatment is unpractical for very low frequency or static deformations because $y_3$ becomes undetermined. Longman (1963) solved this problem by assuming the so-called Adams-Williamson condition,

$$\text{(A-20)} \quad d\rho/dr + \rho^2 g/\lambda = 0$$

which asserts that the density stratification must be adiabatic. In fact, this condition is satisfied very well in the earth's liquid core. Unfortunately, the solutions given by Longman and several other authors after him did not include details about the displacements in the core. In order to attain the complete solutions we shall deduce the equations of motion in a liquid medium, where the Adams-Williamson condition is assumed to be valid, in another way.

Let us differentiate equation (A-19) and substitute $dy_1/dr$, $dy_2/dr$, $dy_5/dr$ by the known equations in (A-17), $d\rho/dr$ by the Adams-Williamson condition (A-20), and $dg/dr$ by relation

$$\text{(A-21)} \quad dg/dr = -2g/r + 4\pi G\rho .$$

Then we obtain a new equation concerning $dy_3/dr$

$$dy_3/dr = y_1/r - y_3/r.$$

Using a new set of notations to distinguish the liquid case from the solid case

$$\text{(A-22)} \quad \mathbf{x} = \begin{bmatrix} x_1 \\ x_2 \\ x_3 \\ x_4 \end{bmatrix} = \begin{bmatrix} y_1 \\ y_3 \\ y_5 \\ y_6 \end{bmatrix},$$

the equation system in a liquid layer reduces to

$$\text{(A-23)} \quad d\mathbf{x}/dr = \mathbf{B}\mathbf{x}$$

where $\mathbf{B}$ is a 4×4-matrix with the coefficients

$$B_{11} = \rho g/\lambda - 2/r$$

$$B_{12} = n(n+1)/r - \rho\omega^2 r/\lambda$$

$$B_{13} = -\rho/\lambda$$

$$B_{21} = 1/r$$

$$B_{22} = -1/r$$

$$B_{31} = 4\pi G\rho$$

$$B_{34} = 1$$

$$B_{42} = -4\pi G\rho n(n+1)/r$$

$$B_{43} = n(n+1)/r^2$$

$$B_{44} = -2/r$$

all other $B_{ij} = 0$.

The variable $y_2$ will be determined directly by relation (A-19). It should be noted that the system of differential equations (A-23) is valid for dynamical deformations ($\omega \neq 0$). In this case it can provide the complete solutions. For statical deformations ($\omega = 0$), the solutions determined by system (A-23) should be understood as the limit ($\omega \rightarrow 0$) of the dynamical solutions.

As in the case of toroidal deformations, equations (A-17) and (A-23) are also poorly conditioned for numerical computation. For solids, this difficulty can be surmounted by making the transformations

$$\text{(A-24)} \qquad \tilde{\mathbf{y}} = r^{1/2} \begin{bmatrix} y_1 \\ r y_2/(n+1) \\ n y_3 \\ r y_4 \\ y_5 \\ r y_6 + (n+1) y_5 \end{bmatrix}$$

$$r = \exp[z/(n+1)].$$

The transformed coefficient matrix $\tilde{\mathbf{A}}$ has the elements

$$\tilde{A}_{11} = (-3\lambda + \mu)/[(n+1)(\lambda + 2\mu)]$$

$$\tilde{A}_{12} = 1/(\lambda + 2\mu)$$

$$\tilde{A}_{13} = \lambda/[(\lambda + 2\mu)]$$

$$\tilde{A}_{21} = [4\mu(3\lambda + 2\mu)/(\lambda + 2\mu) - 4\rho g r - \rho\omega^2 r^2]/(n+1)^2$$

$$\tilde{A}_{22} = -\tilde{A}_{11}$$

$$\tilde{A}_{23} = [\rho g r - 2\mu(3\lambda + 2\mu)/(\lambda + 2\mu)]/(n+1)$$

$$\tilde{A}_{24} = n/(n+1)$$

$$\tilde{A}_{25} = \rho r/(n+1)$$

$$\tilde{A}_{26} = -\rho r/(n+1)^2$$

$\tilde{A}_{31} = -\tilde{A}_{24}$

$\tilde{A}_{33} = 3/[2(n+1)]$

$\tilde{A}_{34} = n/[\mu(n+1)]$

$\tilde{A}_{41} = \tilde{A}_{23}$

$\tilde{A}_{42} = -\tilde{A}_{13}$

$\tilde{A}_{43} = 2\mu\{2(\lambda+\mu)/(\lambda+2\mu) - 1/[n(n+1)]\} - \rho\omega^2 r^2/[n(n+1)]$

$\tilde{A}_{44} = -\tilde{A}_{33}$

$\tilde{A}_{45} = -\rho r/(n+1)$

$\tilde{A}_{51} = 4\pi G\rho r/(n+1)$

$\tilde{A}_{55} = -(n+1/2)/(n+1)$

$\tilde{A}_{56} = 1/(n+1)$

$\tilde{A}_{61} = 4\pi G\rho r$

$\tilde{A}_{63} = -4\pi G\rho r$

$\tilde{A}_{66} = -\tilde{A}_{55}$

all other $\tilde{A}_{ij} = 0$.

For liquids, the following transformations may be suggested

(A-25)
$$\tilde{\mathbf{x}} = r^{1/2}\begin{bmatrix} 4\pi G\rho r x_1 \\ n r x_2 \\ n x_3 \\ r x_4 + (n+1)x_3 \end{bmatrix}$$

$r = \exp[z/(n+1)]$.

Then the transformed coefficient matrix $\tilde{\mathbf{B}}$ has the elements

$\tilde{B}_{11} = -1/(2n + 2)$

$\tilde{B}_{12} = 4\pi G\rho\{1 - \rho\omega^2 r^2/[\lambda n(n + 1)]\}$

$\tilde{B}_{13} = -4\pi G\rho^2 r^2/[\lambda n(n + 1)]$

$\tilde{B}_{21} = n/[4\pi G\rho(n + 1)]$

$\tilde{B}_{22} = -\tilde{B}_{11}$

$\tilde{B}_{31} = n/(n + 1)$

$\tilde{B}_{33} = -(2n + 1)/(2n + 2)$

$\tilde{B}_{34} = \tilde{B}_{31}$

$\tilde{B}_{41} = 1$

$\tilde{B}_{42} = -4\pi G\rho$

$\tilde{B}_{44} = -\tilde{B}_{33}$

all other $\tilde{B}_{ij} = 0$.

For spheroidal solutions we need to determine the starting vectors near the center, where an uniform sphere will be considered. The relatively more simple case is of degree n = 0. In this case we can solve the equations analytically. In an uniform solid sphere,

$$\text{(A-26)} \quad \begin{cases} y_1 = C \cdot j_1(kr) \\ y_2 = (\lambda + 2\mu)dy_1/dr + 2\lambda y_1/r \end{cases}$$

where C is a non-zero constant, $j_1$ is the spherical Bessel function of first degree, and

$$\text{(A-27)} \quad k^2 = \rho(\omega^2 + 4f)/(\lambda + 2\mu), \quad f = 4\pi G\rho/3.$$

If an incompressible starting sphere is considered, the solution reduces simply to

$$\text{(A-28)} \quad \begin{cases} y_1 = 0, \\ y_2 = C. \end{cases}$$

The non-zero constant C, as in the case of the toroidal solutions, will be determined by the surface conditions. The starting value of the third variable $y_5$ at first may be set equal to zero. After integration from the starting radius to the surface, we make $y_5$ vanish at the surface by subtracting its surface value from it.

The boundary conditions for this case are that all variables $y_1$, $y_2$ and $y_5$ are continuous everywhere.

The starting vectors for degree n » 0 are much more complicated. Here we suggest series solutions, which are convenient for numerical programming. In an uniform solid sphere, the transformed equation system can be rewitten as

$$\text{(A-28)} \quad r d\tilde{\mathbf{y}}/dr = (n+1)\tilde{\mathbf{A}}\tilde{\mathbf{y}} = \{\tilde{\mathbf{A}}^{(0)} + r\tilde{\mathbf{A}}^{(1)} + r^2\tilde{\mathbf{A}}^{(2)}\}\tilde{\mathbf{y}}$$

where $\tilde{\mathbf{A}}^{(i)}$ ( i = 0, 1, 2 ) are $6\times6$-matrices with constant elements. Substituting the series solutions of the form

$$\text{(A-29)} \quad \tilde{\mathbf{y}} = r^{\alpha} \sum_{j=0}^{\infty} r^j \mathbf{b}^{(j)}$$

in equations (A-28), where $\alpha$ and $\mathbf{b}^{(j)}$ are constant exponent and vectors to be determined, respectively, we obtain a series of algebraical equations:

$$\{\alpha \mathbf{I} - \tilde{\mathbf{A}}^{(0)}\}\mathbf{b}^{(0)} = 0$$

$$\{(\alpha+1)\mathbf{I} - \tilde{\mathbf{A}}^{(0)}\}\mathbf{b}^{(1)} = \tilde{\mathbf{A}}^{(1)}\mathbf{b}^{(0)}$$

$$\{(\alpha+2)\mathbf{I} - \tilde{\mathbf{A}}^{(0)}\}\mathbf{b}^{(2)} = \tilde{\mathbf{A}}^{(1)}\mathbf{b}^{(1)} + \tilde{\mathbf{A}}^{(2)}\mathbf{b}^{(0)}$$

...

$$\{(\alpha+j)\mathbf{I} - \tilde{\mathbf{A}}^{(0)}\}\mathbf{b}^{(j)} = \tilde{\mathbf{A}}^{(1)}\mathbf{b}^{(j-1)} + \tilde{\mathbf{A}}^{(2)}\mathbf{b}^{(j-2)}, \quad j \geq 2.$$

where $\mathbf{I}$ is the unit matrix. Now the determination of the starting vectors is converted to an eigenvalue problem of algebra. There are three eigenvalues whose

**corresponding eigenvectors are regular at the center. All of them can be determined analytically and are given as follows:**

(A-30) $\alpha_1 = n - 1/2$:

$$\mathbf{b}^{(0)} = \left\{ 0, \frac{2(n-1)\mu}{(n+1)}, n, 2(n-1)\mu, 0, 0 \right\}^T;$$

$$\mathbf{b}^{(1)} = \left\{ 0, 0, 0, 0, nf - \omega^2, 2n(n-1)f - (2n+1)\omega^2 \right\}^T;$$

$$\mathbf{b}^{(j)} = \mathbf{0}, \quad \text{for } j \geq 2.$$

(A-31) $\alpha_2 = n + 1/2$:

$$\mathbf{b}^{(0)} = \left\{ 0, 0, 0, 0, 1, 2n+1 \right\}^T;$$

$$\mathbf{b}^{(1)} = \frac{n\rho}{(n+1)[n(\lambda+\mu) - 2\mu]} \left\{ 0, -\lambda, 1, \mu, 0, 0 \right\}^T;$$

$$\mathbf{b}^{(j)} = \left\{ (n+j+1/2)\mathbf{I} - \tilde{\mathbf{A}}^{(0)} \right\}^{-1} \left\{ \tilde{\mathbf{A}}^{(1)} \mathbf{b}^{(j-1)} + \tilde{\mathbf{A}}^{(2)} \mathbf{b}^{(j-2)} \right\}, \text{ for } j \geq 2.$$

(A-32) $\alpha_3 = n + 3/2$:

$$\mathbf{b}^{(0)} = \begin{bmatrix} (n+1)[n(\lambda+\mu) - 2\mu] \\ 2\mu[n(n-1)(\lambda+\mu) - (3\lambda+2\mu) \\ n[(n+1)(\lambda+\mu) + 2(\lambda+2\mu)] \\ 2\mu[(n+1)^2(\lambda+\mu) - (\lambda+2\mu)] \\ 0 \\ 0 \end{bmatrix};$$

$$\mathbf{b}^{(1)} = \begin{bmatrix} 0 \\ 0 \\ 0 \\ 0 \\ \frac{3(n+1)f}{(2n+3)}[2n(\lambda+\mu) + (2n-1)\mu] \\ -3(n+1)f[n(\lambda+\mu) + (2n+1)\mu] \end{bmatrix};$$

$$\mathbf{b}^{(j)} = \left\{ (n+j+3/2)\mathbf{I} - \tilde{\mathbf{A}}^{(0)} \right\}^{-1} \left\{ \tilde{\mathbf{A}}^{(1)} \mathbf{b}^{(j-1)} + \tilde{\mathbf{A}}^{(2)} \mathbf{b}^{(j-2)} \right\}, \text{ for } j \geq 2.$$

where $f = 4\pi G\rho/3$. Thus the three starting vectors, which are mutually linearly independent, are

$$\tilde{\mathbf{y}}^{(i)} = r^{\alpha_i} \sum_{j=0}^{\infty} r^j \mathbf{b}^{(j)}(\alpha_i), \quad i = 1, 2, 3. \tag{A-33}$$

For an incompressible liquid sphere there are two independent solution vectors. They can be determined analytically:

$$\mathbf{x}^{(1)} = \begin{bmatrix} n r^{n-1} \\ r^{n-1} \\ 2nf r^n \\ n(2n - 3)f r^{n-1} \end{bmatrix} \tag{A-34}$$

and

$$\mathbf{x}^{(2)} = \begin{bmatrix} n r^{n-1} \\ r^{n-1} \\ 3f r^n \\ 0 \end{bmatrix}. \tag{A-35}$$

The starting radius $r_o$ should be chosen sufficiently near the center so that the uniform layer solution is a suitably accurate approximation to that for the real layer. However, if $r_o$ is too small, the computational effort will increase quickly and the numerical accuracy will be affected. In fact, the solution decreases roughly exponentially with depth and degree of the spherical harmonics. For example, if $n \geq 10$, the whole core does not play any part in the results. Additionally, for a well conditioned equation system the initial errors in the starting values can be restrained during integration. Empirically we can choose

$$r_o / a = \text{const.}\ \exp\left( \frac{\ln \varepsilon}{n + 1} \right) \tag{A-36}$$

where a is the earth's radius, $\varepsilon$ the required accuracy of integration, and the constant is of order 1.

All the variables are known to be continuous through any welded boundary in solids. In the following, we discuss the boundary conditions at the assumed frictionless boundaries from solid to liquid, from liquid to liquid with density discontinuity, and from liquid to solid, respectively.

( a ) *Solid to liquid*: The three vectors of the solid side will be, at first, linearly combined to two independent vectors by the condition $y_4 = 0$, and then continuously brought to the liquid side, except for the component $y_3$. The jump of $y_3$ will be determined by relation (A-19), if $\omega \neq 0$. In the case $\omega = 0$, relation (A-19) provides an additional restriction, which is used to linearly combine the two independent vectors into one single vector. The second vector of the liquid side will be obtained by inserting a new non-zero constant for the jump of $y_3$. This process can be displayed schematically as

$$\left.\begin{matrix}\mathbf{y}^{(1)}\\ \mathbf{y}^{(2)}\\ \mathbf{y}^{(3)}\end{matrix}\right|_{r=b^-} \overset{y_4=0}{=\!=\!=\!=\!\Rightarrow} \begin{cases} \left.\begin{matrix}\mathbf{y}^{(1)}\\ \mathbf{y}^{(2)}\end{matrix}\right|_{r=b^+}, \; y_3^+ = \dfrac{1}{b\omega^2}\left\{[gy_1 - y_5]^+ - y_2^-/\rho^+\right\}; \\[2ex] \left.\begin{matrix}\mathbf{y}^{(1)}\\ \mathbf{y}^{(2)}\end{matrix}\right|_{r=b^-} \begin{matrix} \overset{[gy_1 - y_5]^+ = y_2^-/\rho^+}{=\!=\!=\!=\!=\!=\!=\!=\!=\!\Rightarrow} \mathbf{y}^{(1)} \\ y_3^+ = \text{const.} =\!=\!=\!\Rightarrow \mathbf{y}^{(2)} \end{matrix}\Bigg|_{r=b^+}, \quad \omega = 0. \end{cases}$$

( b ) *Liquid to liquid with density discontinuity*: If $\omega \neq 0$, the two independent vectors will be continuously brought through the boundary except for the component $y_3$. The jump of this component can be determined by the condition of continuity of $y_2$:

$$[y_2]_-^+ = [\rho(gy_1 - y_5 - r\omega^2 y_3)]_-^+ = 0 .$$

For static deformations ($\omega = 0$), this condition leads to vanishing of $y_2$ at the boundary and constitutes a restriction, which again is used to combine the two vectors into a single one. The second independent vector is constructed by a non-zero jump of $y_3$, which will be determined finally by the surface conditions. Schematically, this procedure may be displayed as:

$$\left.\begin{matrix}\mathbf{y}^{(1)}\\ \mathbf{y}^{(2)}\end{matrix}\right|_{r=b^-} \begin{cases} =\!=\!\Rightarrow \left.\begin{matrix}\mathbf{y}^{(1)}\\ \mathbf{y}^{(2)}\end{matrix}\right|_{r=b^+}, \; y_3^+ = \beta y_3^- + \dfrac{1}{b\omega^2}(1-\beta)(gy_1 - y_5); \\[2ex] \overset{y_2=0}{=\!=\!=\!=\!\Rightarrow} \mathbf{y}^{(1)}\Big|_{r=b^-} \begin{matrix} =\!=\!=\!=\!=\!=\!=\!=\!\Rightarrow \mathbf{y}^{(1)} \\ y_3^+ = \text{const.} =\!=\!=\!\Rightarrow \mathbf{y}^{(2)} \end{matrix}\Bigg|_{r=b^+}, \quad \text{for } \omega = 0. \end{cases}$$

where $\beta = \rho^-/\rho^+$.

( c ) *Liquid to solid*: The two independent vectors, except for the component $y_3$

will be brought continuously from the liquid side to the solid side. Inserting a new non-zero constant for the jump of $y_3$ leads to the construction of the third independent vector. The component $y_3$ in the first and second vectors can be set equal to zero. We show this process schematically as

$$
\left.\begin{matrix} \mathbf{y}^{(1)} \\ \mathbf{y}^{(2)} \end{matrix}\right|_{r=b^-}
\begin{matrix} \Longrightarrow \\ y_3^+ = \text{const.} \Longrightarrow \end{matrix}
\left.\begin{matrix} \mathbf{y}^{(1)} \\ \mathbf{y}^{(2)} \\ \mathbf{y}^{(3)} \end{matrix}\right|_{r=b^+} .
$$

## Appendix B. Integration of the Triple Products of Spherical Harmonics

The complex normalized surface spherical harmonics have been defined in Chapter 4:

$$\text{(B-1)} \qquad Y_{nm}(\vartheta, \lambda) = C_n^m P_n^m(\cos\vartheta)\exp(im\lambda)$$

$$C_n^m = \left\{\frac{(2n+1)(n-m)!}{4\pi(n+m)!}\right\}^{1/2}$$

for $n = 0, 1, 2, \cdots$ ; and $m = -n, \cdots, n$.

where $\vartheta$ is the geocentric co-latitude, $\lambda$ the geocentric longitude and $P_n^m(\cos\vartheta)$ the associated Legendre functions

$$\text{(B-2)} \qquad P_n^m(x) = \frac{(1-x^2)^{m/2}}{2^n n!}\frac{d^{n+m}}{dx^{n+m}}(x^2-1)^n$$

$$= \frac{(1-x^2)^{m/2}}{2^n}\sum_{k=0}^{[(n-m)/2]}\frac{(-1)^k(2n-2k)!\,x^{n-m-2k}}{k!(n-k)!(n-m-2k)!},$$

$$P_n^{-m}(x) = (-1)^m\frac{(n-m)!}{(n+m)!}P_n^m(x) \quad . \quad m = 0, 1, \cdots, n$$

where $[(n-m)/2]$ denotes the integer part of $(n-m)/2$. The normalization condition is

$$\text{(B-3)} \qquad \int_0^{2\pi} d\lambda \int_0^{\pi} d\vartheta \sin\vartheta\, Y_{nm}(\vartheta, \lambda)\cdot Y_{lk}^*(\vartheta, \lambda) = \delta_{n,l}\cdot\delta_{m,k}$$

where the asterisk denotes the complex conjugate.

In Chapter 3 it has been shown that the tidal anomalies at the earth's surface can be evaluated by integrals over the whole earth, whose kernel functions are expressed in terms of triple products of the parameter perturbations, the undisturbed tidal deformations and the so-called auxiliary solutions, or their derivatives. Using the expansions in spherical harmonics, these volume integrals may be separated into series of products of radial and surface integrals. The radial integrals have to be carried out from the earth's origin to the surface and, in general, have to be calculated numerically. The surface integrals, which consist only of triple products of surface spherical harmonics or their derivatives, will be carried out over an unit sphere surface and, as will be seen below, can be solved analytically.

One fundamental type of the surface integrals is

(B-4) $$A(1,2,3) = \int_0^{2\pi} d\lambda \int_0^{\pi} d\vartheta \sin\vartheta\, Y_1(\vartheta,\lambda) Y_2(\vartheta,\lambda) Y_3(\vartheta,\lambda)$$

where we write the index group $(n_1, m_1)$ simply by 1, and so on. After integrating by $\lambda$ we obtain

(B-5) $$A(1,2,3) = 2\pi C_{n_1}^{m_1} C_{n_2}^{m_2} C_{n_3}^{m_3} \delta_{-m_2,\, m_1+m_3} \cdot F(1,2,3)$$

where

(B-6) $$F(1,2,3) = \int_0^{\pi} d\vartheta \sin\vartheta\, P_{n_1}^{m_1}(\cos\vartheta) P_{n_2}^{m_2}(\cos\vartheta) P_{n_3}^{m_3}(\cos\vartheta).$$

Without loss of generality, we can suppose that all the orders $m_1$, $m_2$ and $m_3$ are not negative, since $P_n^{-m}(x)$ is proportional to $P_n^m(x)$. Substituting the series form (B-2) for the associated Legendre functions and using the following formula:

(B-7) $$\int_0^{\pi} d\vartheta \sin\vartheta \cdot \sin^m\vartheta \cdot \cos^n\vartheta = [1 + (-1)^n] \frac{m!!\,(n-1)!!}{(n+m+1)!!}$$

$$m = 0, 2, 4, \cdots$$

where $(2k)!! = 2\cdot 4\cdot 6 \cdots (2k)$, $(2k+1)!! = 1\cdot 3\cdot 5 \cdots (2k+1)$ and we define that $0!! = (-1)!! = 1$, $F(1, 2, 3)$ can be expressed as

(B-8) $$F(1,2,3) = \sum_{k_1} \sum_{k_2} \sum_{k_3} p_1 p_2 p_3 \cdot [1 + (-1)^{N-M}] \frac{M!!\,(N-M-2K-1)!!}{(N-2K+1)!!}$$

where

$$N = n_1 + n_2 + n_3$$

$$M = m_1 + m_2 + m_3 \quad (= 0, 2, 4, \cdots)$$

$$K = k_1 + k_2 + k_3$$

$$p_s = (-1)^{k_s} \frac{(2n_s - 2k_s)!}{2^{n_s} k_s!\,(n_s - k_s)!\,(n_s - m_s - 2k_s)!}, \quad s = 1, 2, 3$$

and the summation indices $k_s$ ( $s = 1, 2, 3$ ) are taken from 0 to $[\,(n_s - m_s)/2\,]$.

The other fundamental type of surface integrals is

$$\text{(B-9)} \qquad B(1, 2, 3) = \int_0^{2\pi} d\lambda \int_0^{\pi} d\vartheta \sin\vartheta \, Y_1(\vartheta, \lambda) \, i\mathbf{e}_r \cdot [\nabla_1 Y_2(\vartheta, \lambda) \times \nabla_1 Y_3(\vartheta, \lambda)]$$

where

$$\nabla_1 = \mathbf{e}_\vartheta \frac{\partial}{\partial\vartheta} + \mathbf{e}_\lambda \frac{1}{\sin\vartheta} \frac{\partial}{\partial\lambda}.$$

After taking the surface gradients of spherical harmonics and integrating by $\lambda$, integral (B-9) reduces to

$$\text{(B-10)} \qquad B(1, 2, 3) = 2\pi C_{n_1}^{m_1} C_{n_2}^{m_2} C_{n_3}^{m_3} \delta_{-m_2, m_1+m_3} \int_{-1}^{1} dx \, P_{n_1}^{m_1}(x)$$

$$\times \left\{ -m_2 P_{n_2}^{m_2}(x) \dot{P}_{n_3}^{m_3}(x) + m_3 \dot{P}_{n_2}^{m_2}(x) P_{n_3}^{m_3}(x) \right\}.$$

The terms in the brackets can be substituted by

$$(m_1 + m_3) P_{n_2}^{m_2}(x) \dot{P}_{n_3}^{m_3}(x) + m_3 \dot{P}_{n_2}^{m_2}(x) P_{n_3}^{m_3}(x)$$

$$= m_1 P_{n_2}^{m_2}(x) \dot{P}_{n_3}^{m_3}(x) + m_3 \frac{d}{dx} \left( P_{n_2}^{m_2}(x) P_{n_3}^{m_3}(x) \right).$$

By integrating the term $P_{n_1}^{m_1}(x) \frac{d}{dx} \left( P_{n_2}^{m_2}(x) P_{n_3}^{m_3}(x) \right)$ by part and noting that $P_{n_1}^{m_1}(x) P_{n_2}^{m_2}(x) P_{n_3}^{m_3}(x) = 0$ for $x = \pm 1$, if one of the indexes $m_1$, $m_2$ and $m_3$ is non-zero, it follows that

$$B(1, 2, 3) = - B(2, 1, 3).$$

Similarly, it can be shown that

$$B(1, 2, 3) = - B(3, 2, 1) = B(2, 3, 1) = - B(1, 3, 2).$$

Thus, the integral B has the following recurrence property:

(B-11) $$B(i, j, k) = B(1, 2, 3)\,\varepsilon_{ijk}$$

where $\varepsilon_{ijk}$ is called the total asymmetric tensor, or Levi-Civita tensor, and can be expressed by

$$\varepsilon_{ijk} = \mathbf{e}_i \cdot (\mathbf{e}_j \times \mathbf{e}_k)$$

where $\mathbf{e}_1$, $\mathbf{e}_2$ and $\mathbf{e}_3$ are three unit vectors which form a right-handed orthogonal system, and (i, j, k) is any arrangement of the index groups (1, 2, 3). Consequently, B will vanish, if any two of the three harmonics $Y_1$, $Y_2$ and $Y_3$ are identical or zonal.

Using the recurrence formula

$$\dot{P}_n^m = \frac{1}{(2n+1)}\left[(n+1)(n+m)P_{n-1}^m - n(n-m+1)P_{n+1}^m\right]/(1-x^2)$$

the integral in (B-10) can be expressed in terms of the following integrals

(B-12) $$G(1, 2, 3) = \int_{-1}^{1} \frac{dx}{(1-x^2)} P_{n_1}^{m_1}(x)\,P_{n_2}^{m_2}(x)\,P_{n_3}^{m_3}(x),$$

$$(|m_1| + |m_2| + |m_3| \geq 2).$$

Supposing that all the orders $m_1$, $m_2$ and $m_3$ are not negative, its series solution is

(B-13) $$G(1, 2, 3) = \sum_{k_1}\sum_{k_2}\sum_{k_3} p_1\, p_2\, p_3 \cdot [1 + (-1)^{N-M}]\frac{M!!\,(N - M - 2K - 1)!!}{M\cdot(N - 2K - 1)!!}$$

Here all notations are the same as above. Integral (B-10) then reduces to

(B-14) $$B(1,2,3) = 2\pi C_{n_1}^{m_1} C_{n_2}^{m_2} C_{n_3}^{m_3} \delta_{-m_2,\, m_1+m_3} \times$$

$$\times \Big\{ \frac{-m_2}{(2n_3+1)} \Big[ (n_3+1)(n_3+m_3)\, G(n_3-1) - n_3(n_3-m_3+1)\, G(n_3+1) \Big]$$

$$+ \frac{m_3}{(2n_2+1)} \Big[ (n_2+1)(n_2+m_2)\, G(n_2-1) - n_2(n_2-m_2+1)\, G(n_2+1) \Big] \Big\}$$

where $G(n_3-1) = G(n_1, m_1; n_2, m_2; n_3-1, m_3)$ and so on.

In the following, we give expressions of diverse integrals in terms of the fundamental integrals A and B. They are very useful in Chapter 6 for the computationally convenient formularization of the perturbation method. Let us define the solid spherical harmonics as

(B-15) $$w_{nm}(r, \vartheta, \lambda) = r^n Y_{nm}(\vartheta, \lambda)$$

and suppose $f(r)$ being an arbitrary radially dependent function, then the following relations hold:

(B-16) $$\iiint\limits_{r \le a} f \cdot w_1 w_2 w_3 \, dv = A(1,2,3) \int_0^a f(r)\, r^{N+2}\, dr$$

(B-17) $$\iiint\limits_{r \le a} f \cdot w_1 \nabla w_2 \cdot \nabla w_3 \, dv$$

$$= \frac{1}{2}(N+1)(n_2+n_3-n_1)\, A(1,2,3) \int_0^a f(r)\, r^N dr$$

(B-18) $$\iiint\limits_{r \le a} f \cdot w_1 \nabla\nabla w_2 : \nabla\nabla w_3 \, dv$$

$$= \frac{1}{4}(N^2-1)(n_2+n_3-n_1-2)(n_2+n_3-n_1)\, A(1,2,3) \int_0^a f(r)\, r^{N-2} dr$$

(B-19) $$\iiint\limits_{r \le a} f \cdot \nabla\nabla w_1 : (\nabla w_2 \nabla w_3) \, dv$$

$$= \frac{1}{4}(N^2-1)\left[n_1^2 - (n_2-n_3)^2\right] A(1,2,3) \int_0^a f(r)\, r^{N-2} dr$$

(B-20) $\iiint\limits_{r\le a} f\cdot w_1\,(\mathbf{r}\times\nabla w_2)\cdot(\mathbf{r}\times\nabla w_3)\,dv$

$$=\left[\frac{1}{2}(N+1)(n_2+n_3-n_1)-n_2n_3\right]A(1,\,2,\,3)\int_0^a f(r)\,r^{N+2}\,dr$$

(B-21) $\iiint\limits_{r\le a} f\cdot w_1\,\nabla(\mathbf{r}\times\nabla w_2):[\nabla(\mathbf{r}\times\nabla w_3)]^T\,dv$

$$=\frac{1}{4}(N+1)(n_2+n_3-n_1)[(N+1)(n_2+n_3-n_1)-2n_2n_3]\times$$

$$\times A(1,\,2,\,3)\int_0^a f(r)\,r^N\,dr$$

(B-22) $\iiint\limits_{r\le a} f\cdot w_1\,\nabla(\mathbf{r}\times\nabla w_2):\nabla(\mathbf{r}\times\nabla w_3)\,dv$

$$=\frac{1}{4}(N+1)(n_2+n_3-n_1)[(N+1)(n_2+n_3-n_1)$$

$$-(2n_2+1)(2n_3+1)-1]\,A(1,\,2,\,3)\int_0^a f(r)\,r^N\,dr$$

(B-23) $\iiint\limits_{r\le a} f\,(\mathbf{r}\times\nabla w_1)\cdot\nabla\nabla w_2\cdot(\mathbf{r}\times\nabla w_3)\,dv$

$$=\frac{1}{4}(N+1)[n_2^2-(n_1-n_3)^2](2n_2-N-1)\,A(1,\,2,\,3)\int_0^a f(r)\,r^N\,dr$$

(B-24) $\iiint\limits_{r\le a} f\cdot w_1\,\mathbf{r}\cdot(\nabla w_2\times\nabla w_3)\,dv$

$$=-i\,B(1,\,2,\,3)\int_0^a f(r)\,r^{N+1}\,dr$$

(B-25) $\iiint\limits_{r\le a} f\,\nabla w_1\cdot\nabla\nabla w_2\cdot(\mathbf{r}\times\nabla w_3)\,dv$

$$=i\,\frac{1}{2}N(n_1+n_2-n_3-1)\,B(1,\,2,\,3)\int_0^a f(r)\,r^{N-1}\,dr$$

(B-26) $$\iiint_{r\le a} f\, w_1 \nabla\nabla w_2 : \nabla(\mathbf{r}\times\nabla w_3)\, dv$$

$$= i\frac{1}{2} N(n_2+n_3-n_1-1) B(1,2,3) \int_0^a f(r) r^{N-1} dr$$

(B-27) $$\iiint_{r\le a} f\, \nabla w_1 \cdot (\nabla w_2 \times \nabla w_3)\, dv$$

$$= -iN\cdot B(1,2,3) \int_0^a f(r) r^{N-1} dr .$$

As an example, here we prove the last relation (B-27). Note that

$$\nabla w = \mathbf{e}_r \frac{\partial w}{\partial r} + \frac{1}{r}\nabla_1 w = r^{(n-1)}\left\{ \mathbf{e}_r nY + \nabla_1 Y \right\}$$

$$\mathbf{e}_r \cdot (\mathbf{e}_r \times \nabla_1 w) = \nabla_1 w_1 \cdot (\nabla_1 w_2 \times \nabla_1 w_3) = 0$$

we obtain

$$\nabla w_1 \cdot (\nabla w_2 \times \nabla w_3)$$

$$= r^{(N-1)}\left\{ n_1 Y_1 \mathbf{e}_r \cdot (\nabla_1 Y_2 \times \nabla_1 Y_3) + n_2 Y_2 \mathbf{e}_r \cdot (\nabla_1 Y_3 \times \nabla_1 Y_1) + n_3 Y_3 \mathbf{e}_r \cdot (\nabla_1 Y_1 \times \nabla_1 Y_2) \right\}$$

and the left-hand of relation (B-27) reduces to

$$\iiint_{r\le a} f\, \nabla w_1 \cdot (\nabla w_2 \times \nabla w_3)\, dv$$

$$= -i\left\{ n_1 B(1,2,3) + n_2 B(2,3,1) + n_3 B(3,1,2) \right\} \int_0^a f(r) r^{N-1} dr .$$

Using the recurrence property of integral B, (B-27) is verified.

In Chapter 5 and Appendix C, one of the three spherical harmonics in the triple products takes the complex conjugate form. Instead of (B-4) and (B-9), the two fundamental types of the surface integrals $\tilde{A}$ and $\tilde{B}$ are defined by

(B-28) $$\tilde{A}(n_o, m_o; n, m; l, k) = \int_0^{2\pi} d\lambda \int_0^{\pi} d\vartheta \sin\vartheta \, Y_{n_o m_o} Y^*_{nm} Y_{lk}$$

(B-29) $$\tilde{B}(n_o, m_o; n, m; l, k)$$

$$= \int_0^{2\pi} d\lambda \int_0^{\pi} d\vartheta \sin\vartheta \, Y_{n_o m_o} i \, \mathbf{e}_r \cdot (\nabla_1 Y^*_{nm} \times \nabla_1 Y_{lk}).$$

Because

$$Y^*_{nm} = (-1)^m Y_{n,-m}$$

we obtain immediately

(B-30) $$\tilde{A}(n_o, m_o; n, m; l, k) = (-1)^m A(n_o, m_o; n, -m; l, k)$$

and

(B-31) $$\tilde{B}(n_o, m_o; n, m; l, k) = (-1)^m B(n_o, m_o; n, -m; l, k).$$

## Appendix C. Evaluation of Integrals $\Gamma^j_{nm}$

It has been shown in Chapter 3 and Chapter 5 that the expansion coefficients of the tidal anomalies at the earth's surface may be evaluated by integrals $\Gamma^j_{nm}$ whose kernel functions depend only upon the parameter perturbations, the undisturbed tidal deformations and the auxiliary solutions of the undisturbed system. Using expansions in spherical harmonics these integrals can be solved quasi-analytically. In this appendix we give all the formulae for evaluating integrals $\Gamma^j_{nm}$ with respect to different parameter perturbations.

According to the notation defined in the chapters, indexes $n_o$ and $m_o$ are used to denote the degree and order of the undisturbed solution, indexes n and m for the auxiliary solution, and indices l and k for the parameter perturbations. The coefficients $\tilde{A}$ and $\tilde{B}$ with index arguments ( $n_o$, $m_o$; n, m; l, k ) are given in Appendix A.

### C.1 Inertial forces

Here we consider deviations from the static deformations of the non-rotating earth:

$$\text{(C-1)} \qquad \Gamma^j_{nm}(\mathbf{\Omega}, \omega) = \iiint\limits_{r\le a} \rho^o \left\{ [\, \omega^2 \mathbf{u}^o - i2\omega \mathbf{\Omega} \times \mathbf{u}^o \,] \cdot \mathbf{u}^j + \nabla^2 \Psi \cdot u^o_r u^j_r \right\} d\upsilon$$

The last term is actually a part of the indirect effect of the perturbation of the gravitational field due to the earth's rotation. We include it here for the reason that it is simply related to the angular velocity $\Omega$ of the earth's rotation by the Poisson's equation (see below). Using the expansions in spherical harmonics of both the undisturbed and the auxiliary solutions in Chapter 5, and noting that

$$\mathbf{\Omega} = \Omega \mathbf{e}_z = \Omega(\mathbf{e}_r \cos\vartheta - \mathbf{e}_\vartheta \sin\vartheta) = \sqrt{4\pi/3} \cdot \Omega \nabla w_{10}(r, \vartheta, \lambda)$$

$$\nabla^2 \Psi = 2\Omega^2$$

where $w_{10}$ is the normalized solid spherical harmonic (s. App. B), integral (C-1) can be deduced to a series form, in which only one-dimensional radial integrals are to be evaluated:

$$\text{(C-2)} \qquad \Gamma^j_{nm}(\mathbf{\Omega}, \omega) = \sum_{l=0}^{1} i^l \Big\{ \tilde{A}(n_o, m_o; n, m; l, 0) \int_0^a \rho^o r^2 K^j_{\Omega,\omega}(n_o, n, l, r)\, dr + i\tilde{B}(n_o, m_o; n, m; l, 0) \int_0^a \rho^o r^2 G^j_{\Omega,\omega}(n_o, n, l, r)\, dr \Big\}$$

where

( 1 ) for l = 0

$$K^{j}_{\Omega,\omega} = \sqrt{4\pi}\ \Big\{ 2\Omega^{2} H_{n_o} H^{j}_{n} + \omega^{2}[ H_{n_o} H^{j}_{n} + $$
$$+ \frac{1}{2}( n_o( n_o + 1) + n( n + 1))( T_{n_o} T^{j}_{n} + W_{n_o} W^{j}_{n})] \Big\},$$

$G_{\Omega,\omega}$ has no meaning, because coefficient $\tilde{B}$ vanishes in this case,

( 2 ) for l = 1

$$K^{j}_{\Omega,\omega} = -\sqrt{16\pi/3}\cdot\Omega\omega\Big\{ n( H_{n_o} - n_o T_{n_o}) W^{j}_{n} - n_o( H^{j}_{n} - n T^{j}_{n}) W_{n_o}$$
$$+ \frac{1}{2}( n_o + n + l + 1)[ ( n_o + l - n) W_{n_o} H^{j}_{n} - ( n + l - n_o) H_{n_o} W^{j}_{n}$$
$$+ ( n_o + n - l)( T_{n_o} W^{j}_{n} - W_{n_o} T^{j}_{n})] \Big\}$$

$$G^{j}_{\Omega,\omega} = \sqrt{16\pi/3}\cdot\Omega\omega( H_{n_o} T^{j}_{n} + T_{n_o} H^{j}_{n} + T_{n_o} T^{j}_{n} + W_{n_o} W^{j}_{n}).$$

C.2 Heterogeneity and hexagonal anisotropy of elastic moduli

These parameter perturbations have been expressed in terms of spherical harmonics in Chapter 4. The deviations of tidal deformations from that in the laterally homogeneous, isotropic and perfectly elastic case can be described by the integral

$$\text{(C-3)}\quad \Gamma^{j}_{nm}(\delta\mathbf{A}) = -\iiint_{r\le a} \delta_{\mathbf{A}}\boldsymbol{\sigma} : \nabla\mathbf{u}^{j}\, d\upsilon$$
$$= \sum_{f}\sum_{l}\sum_{k} \Big\{ \tilde{A}( n_o, m_o; n, m; l, k) \int_0^a f_{lk} K^{j}_{f}( n_o, n, l, r) dr$$
$$+ i\tilde{B}( n_o, m_o; n, m; l, k ) \int_0^a f_{lk} G^{j}_{f}( n_o, n, l, r) dr \Big\}$$

where f denotes the inhomogeneous or inelastic perturbations $\delta\lambda$ and $\delta\mu$, and the anisotropic perturbations $\delta\lambda_{\perp}$, $\delta\lambda_{//}$, $\delta\mu_{\perp}$, $\delta\mu_{//}$ and $\delta\mu_{\times}$, respectively. The kernel functions K and G are given as

$$K^j_\lambda = -[r\dot{H}_{n_o} + 2H_{n_o} - n_o(n_o+1)T_{n_o}][r\dot{H}^j_n + 2H^j_n - n(n+1)T^j_n]$$

$$G^j_\lambda = 0$$

$$\begin{aligned} K^j_\mu = & -2r^2\dot{H}_{n_o}\dot{H}^j_n - 2H_{n_o}(H^j_n - n(n+1)T^j_n) - \\ & -2H^j_n(H_{n_o} - n_o(n_o+1)T_{n_o}) + n_o n(H_{n_o} + r\dot{T}_{n_o} - T_{n_o})\times \\ & \times(H^j_n + r\dot{T}^j_n - T^j_n) - \frac{1}{2}(n_o + n + l + 1)(n_o + n - l)\times \\ & \times[(H_{n_o} + r\dot{T}_{n_o} - T_{n_o})(H^j_n + r\dot{T}^j_n - T^j_n) - 4(n_o - 1)(n-1)T_{n_o}T^j_n] \\ & - \frac{1}{2}[(n_o + n + l)^2 - 1](n_o + n - l)(n_o + n - l - 2)T_{n_o}T^j_n \\ & -[\frac{1}{2}(n_o + n + l + 1)(n_o + n - l) - n_o n][(r\dot{W}_{n_o} - W_{n_o})\times \\ & \times(r\dot{W}^j_n - W^j_n) - (n_o - 1)(n-1)W_{n_o}W^j_n] - \frac{1}{2}(n_o + n + l + 1)\times \\ & \times(n_o + n - l)[(n_o + n + l + 1)(n_o + n - l) - \\ & -(3n_o n + n_o + n + 1)]W_{n_o}W^j_n \end{aligned}$$

$$\begin{aligned} G^j_\mu = & (r\dot{W}_{n_o} - W_{n_o})(H^j_n + r\dot{T}^j_n - T^j_n) - \\ & -(H_{n_o} + r\dot{T}_{n_o} - T_{n_o})(r\dot{W}^j_n - W^j_n) + [2(n_o - 1)(n-1) + \\ & +(n_o + n + l)(n_o + n - l - 1)](T_{n_o}W^j_n - W_{n_o}T^j_n) \end{aligned}$$

$$K^j_{\lambda_{//}} = -[2H_{n_o} - n_o(n_o+1)T_{n_o}][2H^j_n - n(n+1)T^j_n]$$

$$G^j_{\lambda_{//}} = 0$$

$$K^j_{\lambda_\perp} = K^j_\lambda - K^j_{\lambda_{//}}$$

$$G^j_{\lambda_\perp} = 0$$

$$K^{j}_{\mu_{\perp}} = -2r^{2}\dot{H}_{n_o}\dot{H}^{j}_{n}$$

$$G^{j}_{\mu_{\perp}} = 0$$

$$K^{j}_{\mu_{\times}} = -\frac{1}{2}[(n_o + n + l + 1)(n_o + n - l) - 2n_o n]\times$$

$$\times(r\dot{W}_{n_o} - W_{n_o})(r\dot{W}^{j}_{n} - W^{j}_{n}) - \frac{1}{2}(n_o + n + l + 1)\times$$

$$\times(n_o + n - l)(H_{n_o} + r\dot{T}_{n_o} - T_{n_o})(H^{j}_{n} + r\dot{T}^{j}_{n} - T^{j}_{n})$$

$$G^{j}_{\mu_{\times}} = (H_{n_o} + r\dot{T}_{n_o} - T_{n_o})(r\dot{W}^{j}_{n} - W^{j}_{n}) -$$

$$-(r\dot{W}_{n_o} - W_{n_o})(H^{j}_{n} + r\dot{T}^{j}_{n} - T^{j}_{n})$$

$$K^{j}_{\mu_{//}} = K^{j}_{\mu} - K^{j}_{\mu_{\perp}} - K^{j}_{\mu_{\times}}$$

$$G^{j}_{\mu_{//}} = G^{j}_{\mu} - G^{j}_{\mu_{\times}}.$$

C.3 Density heterogeneity

For this perturbation we consider the following integral

(C-3) $$\Gamma^{j}_{nm}(\delta\rho) = \iiint_{r \le a} \{\delta\rho[(\nabla\varphi^{o} + \mathbf{u}^{o}\cdot\nabla\nabla V^{o})\cdot\mathbf{u}^{j} + \nabla\varphi^{j}\cdot\mathbf{u}^{o}] + \rho^{o}\nabla^{2}\delta\Phi\, u^{o}_{r}u^{j}_{r}\}\, dv.$$

The last term is actually a part of the indirect effect of the potential perturbation due to the density heterogeneity. We include it here because it is simply related to $\delta\rho$ by the Poisson's equation

$$\nabla^{2}\delta\Phi = -4\pi G\delta\rho.$$

The standard formulation for the density perturbation is

$$\text{(C-4)} \qquad \Gamma^{j}_{nm}(\delta\rho) = \sum_{l}\sum_{k} \Big\{ \tilde{A} \int_0^a \rho_{lk} K^{j}_{\rho}(n_o, n, l, r)\,dr + i\tilde{B} \int_0^a \rho_{lk} G^{j}_{\rho}(n_o, n, l, r)\,dr \Big\}$$

where

$$K^{j}_{\rho} = (2g^{o}r - 8\pi G\rho^{o}r^{2}) H_{n_o} H^{j}_{n} + n_o n g^{o} r (T_{n_o} T^{j}_{n} + W_{n_o} W^{j}_{n}) + r^{2}(H_{n_o} \dot{R}^{j}_{n} + \dot{R}_{n_o} H^{j}_{n}) - n_o n r (T_{n_o} R^{j}_{n} + R_{n_o} T^{j}_{n}) + \frac{1}{2}(n_o + n + l + 1)(n_o + n - l) r [T_{n_o} R^{j}_{n} + R_{n_o} T^{j}_{n} - g^{o}(T_{n_o} T^{j}_{n} + W_{n_o} W^{j}_{n})]$$

$$G^{j}_{\rho} = r(R_{n_o} W^{j}_{n} - W_{n_o} R^{j}_{n}) - g^{o} r (T_{n_o} W^{j}_{n} - W_{n_o} T^{j}_{n})$$

and the index arguments of coefficients $\tilde{A}$ and $\tilde{B}$ are omitted henceforth.

C.4 Spherically asymmetric perturbations of the potential

The effect due to this parameter perturbation is given by the integral

$$\text{(C-5)} \qquad \Gamma^{j}_{nm}(\delta V) = \iiint_{r \le a} \rho^{o} \{ \mathbf{u}^{o} \cdot \nabla\nabla\delta V \cdot \mathbf{u}^{j} - \nabla^{2}\delta V\, u^{o}_{r} u^{j}_{r} \}\, d\upsilon .$$

Note that the subtraction of the last term in the brackets is because it has been included in C.1 and C.3 as a part of the effect of the inertial force and of the density heterogeneity, respectively. In the series formula, the radial integrands depend not only on the expansion coefficients of the potential perturbation, but also on their derivatives:

$$\text{(C-6)} \qquad \Gamma^{j}_{nm}(\delta V) = \sum_{l}\sum_{k} \Big\{ \tilde{A} \int_0^a \rho^{o} [ V_{lk} K^{j}_{V}(n_o, n, l, r) + r\dot{V}_{lk} K^{j}_{\dot{V}}(n_o, n, l, r) ]\,dr + i\tilde{B} \int_0^a \rho^{o} [ V_{lk} G^{j}_{V}(n_o, n, l, r) + r\dot{V}_{lk} G^{j}_{\dot{V}}(n_o, n, l, r) ]\,dr \Big\}$$

where

$$K^{j}_{\dot V} = -(l+2)H_{n_o}H^{j}_{n} + l(H_{n_o} - n_oT_{n_o})(H^{j}_{n} - nT^{j}_{n}) -$$

$$- n_o n[(l+1)T_{n_o}T^{j}_{n} + W_{n_o}W^{j}_{n}] + \frac{1}{2}(n_o + n + l + 1)\times$$

$$\times\left\{(n_o + n - l)(T_{n_o}T^{j}_{n} + W_{n_o}W^{j}_{n}) + (n + l - n_o)H_{n_o}T^{j}_{n} +\right.$$

$$\left. + (n_o + l - n)T_{n_o}H^{j}_{n}\right\}$$

$$G^{j}_{\dot V} = W_{n_o}H^{j}_{n} - H_{n_o}W^{j}_{n} + T_{n_o}W^{j}_{n} - W_{n_o}T^{j}_{n}$$

$$K^{j}_{V} = -lK^{j}_{\dot V} + l(l-1)(H_{n_o} - n_oT_{n_o})(H^{j}_{n} - nT^{j}_{n}) +$$

$$+ \frac{1}{2}(n_o + n + l + 1)\left\{(l-1)[(n + l - n_o)(H_{n_o} - n_oT_{n_o})T^{j}_{n} +\right.$$

$$+ (n_o + l - n)T_{n_o}(H^{j}_{n} - nT^{j}_{n})] + \frac{1}{2}[l^2 - (n_o - n)^2]\times$$

$$\left.\times[(n_o + n + l - 1)T_{n_o}T^{j}_{n} - (n_o + n - l + 1)W_{n_o}W^{j}_{n}]\right\}$$

$$G^{j}_{V} = -lG^{j}_{\dot V} + (l-1)[W_{n_o}(H^{j}_{n} - nT^{j}_{n}) - (H_{n_o} - n_oT_{n_o})W^{j}_{n}]$$

$$+ \frac{1}{2}(n_o + n + l)[(n + l - n_o - 1)W_{n_o}T^{j}_{n} -$$

$$- (n_o + l - n - 1)T_{n_o}W^{j}_{n}].$$

## C.5 Perturbation of the generalized hydrostatic pressure

The integral for the effect due to the perturbation of the generalized pressure is given by

$$\text{(C-7)} \qquad \Gamma^{j}_{nm}(\delta P) = \iiint_{r\le a} \left\{\nabla(\mathbf{u}^{o}\cdot\nabla\delta P) - (\nabla\cdot\mathbf{u}^{o})\nabla\delta P - \mathbf{u}^{o}\cdot\nabla\nabla\delta P\right\}\cdot\mathbf{u}^{j}\, d\upsilon$$

$$= \sum_{l}\sum_{k}\left\{\tilde{A}\int_0^a [P_{lk}K^{j}_{P}(n_o, n, l, r) + r\dot{P}_{lk}K^{j}_{\dot P}(n_o, n, l, r)]dr\right.$$

$$\left. + i\tilde{B}\int_0^a [P_{lk}G^{j}_{P}(n_o, n, l, r) + r\dot{P}_{lk}G^{j}_{\dot P}(n_o, n, l, r)]dr\right\}$$

where

$$K_P^{j\cdot} = -(H_{n_o} - n_o^2 T_{n_o})(H_n^j - nT_n^j) - (H_{n_o} - n_o T_{n_o})[H_n^j - n(n_o + 1)T_n^j]$$

$$+ n_o n W_{n_o} W_n^j + \frac{1}{2}(n_o + n + l + 1)(n_o + n - l)\times$$

$$\times(H_{n_o} T_n^j - T_{n_o} T_n^j - W_{n_o} W_n^j)$$

$$G_P^{j\cdot} = W_{n_o} T_n^j + (H_{n_o} - T_{n_o}) W_n^j$$

$$K_P^j = -l K_P^{j\cdot} + l\{[r\dot{H}_{n_o} - (n_o - 1)H_{n_o}]nT_n^j - [2H_{n_o} + n_o r\dot{T}_{n_o} -$$

$$- n_o(n_o + 1)T_{n_o}]H_n^j\} + \frac{1}{2}(n_o + n + l + 1)\times$$

$$\times\{(n + l - n_o)[n_o W_{n_o} W_n^j - (r\dot{H}_{n_o} + H_{n_o} - n_o^2 T_{n_o})T_n^j] +$$

$$+ l(n_o + n - l)[(H_{n_o} - n_o T_{n_o})T_n^j - W_{n_o} W_n^j] +$$

$$+ (n_o + l - n)(r\dot{T}_{n_o} H_n^j - n(n_o - 1)T_{n_o} T_n^j) +$$

$$+ \frac{1}{2}[n_o^2 - (n - l)^2][(n_o + n + l - 1)T_{n_o} T_n^j +$$

$$+ (n + l - n_o + 1)W_{n_o} W_n^j]\}$$

$$G_P^j = -l G_P^{j\cdot} + r\dot{W}_{n_o} H_n^j + r\dot{H}_{n_o} W_n^j - n_o n W_{n_o} T_n^j + (l + 1)H_{n_o} W_n^j -$$

$$- n_o(n + l)T_{n_o} W_n^j + (n_o + n + l)W_{n_o} T_n^j + \frac{1}{2}(n_o + n + l)\times$$

$$\times[(n_o + n - l - 1)W_{n_o} T_n^j + (n_o + l - n - 1)T_{n_o} W_n^j].$$

C.6 Boundary undulations

The effect due to the boundary (surface) undulations is described by surface integrals. Let us define

(C-8) $$\Gamma^j_{nm,1}(\delta h) = \oiint_{r=r^i} \{\delta h[\mathbf{u}^j \cdot \frac{\partial(\mathbf{e}_r \cdot \boldsymbol{\sigma}^o)}{\partial r} + \frac{\varphi^j}{4\pi G}\frac{\partial}{\partial r}(\frac{\partial \varphi^o}{\partial r} - 4\pi G \rho^o u_r^o)] +$$

$$+ \nabla_1 \delta h \cdot (\rho^o \varphi^j \mathbf{u}^o - \boldsymbol{\sigma}^o \cdot \mathbf{u}^j)\} ds$$

$$\text{(C-9)} \qquad \Gamma^{j}_{nm,2}(\delta h) = -\oiint_{r=r^{i}} \left\{ \delta h \left[ \mathbf{e}_r \cdot \boldsymbol{\sigma}^{j} \cdot \frac{\partial \mathbf{u}^{o}}{\partial r} + \frac{1}{4\pi G} \frac{\partial \varphi^{o}}{\partial r} \left( \frac{\partial \varphi^{j}}{\partial r} - 4\pi G \rho^{o} u^{j}_{r} \right) \right] \right\} ds$$

$$\text{(C-10)} \qquad \Gamma^{j}_{nm,3}(\delta h) = \oiint_{r=r^{i}} \frac{\varphi^{j}}{4\pi G} \delta h \left\{ \frac{\partial^{2} \varphi^{o}}{\partial r^{2}} \Big|^{+} + \frac{\partial \delta \varphi}{\partial r} \Big|^{+} + \frac{(n+1)}{r^{i}} \delta \varphi \Big|^{-} \right\} ds .$$

Using the results given in Chapter 3 and 5, we obtain

$$\text{(C-11)} \qquad \Gamma^{j}_{nm}[\delta h(r^{i})] = \left\{ \Gamma^{j}_{nm,1} + \Gamma^{j}_{nm,2} \right\} \Big|^{+}_{-}$$

for an interior boundary undulation, and

$$\text{(C-12)} \qquad \Gamma^{j}_{nm}[\delta h(a)] = \left\{ -\Gamma^{j}_{nm,1} \Big|_{r^{i}=a^{-}} + \Gamma^{j}_{nm,3} \Big|_{r^{i}=a} \right\}$$

for the surface undulation. The analytical series solutions for integrals (C-8) - (C-10) take the form

$$\text{(C-13)} \qquad \Gamma^{j}_{nm,s}(\delta h) = \sum_{l} \sum_{k} \left\{ \tilde{A} K^{j}_{h,s}(n_o, n, l, r^{i}) + i \tilde{B} G^{j}_{h,s}(n_o, n, l, r^{i}) \right\}$$

$$\text{for } s = 1, 2, 3$$

where

$$K^j_{h,1} = (r^i)^2[\dot{y}_{2,n_o}H^j_n - n_o n(\dot{y}_{4,n_o}T^j_n + \dot{y}_{8,n_o}W^j_n)] +$$

$$+ n_o l[r^i y_{4,n_o}H^j_n - \rho^o r^i T_{n_o}R^j_n - n(n_o - 1)\mu^o W_{n_o}W^j_n] +$$

$$+ 2nl\mu^o(H_{n_o} - n_o^2 T_{n_o})T^j_n + \frac{(r^i)^2}{4\pi G} R^j_n[\dot{y}_{6,n_o} -$$

$$- \frac{(n_o+1)}{r^i}\dot{R}_{n_o} + \frac{(n_o+1)}{(r^i)^2}R_{n_o}] + \frac{1}{2}(n_o + n + l + 1)\times$$

$$\times\{(n_o + n - l)[(r^i)^2(\dot{y}_{4,n_o}T^j_n + \dot{y}_{8,n_o}W^j_n) + l(n_o - 1)\mu^o \times$$

$$\times(2T_{n_o}T^j_n + W_{n_o}W^j_n)] + (n_o + l - n)[\rho^o r^i T_{n_o}R^j_n - r^i y_{4,n_o}H^j_n +$$

$$+ 2(n_o - 1)\mu^o(n_o T_{n_o}T^j_n + nW_{n_o}W^j_n)] - 2(n + l - n_o)\mu^o \times$$

$$\times(H_{n_o} - n_o T_{n_o})T^j_n - \frac{1}{2}(n_o + n + l - 1)[n_o^2 - (n - l)^2]\mu^o \times$$

$$\times(2T_{n_o}T^j_n + W_{n_o}W^j_n)\}$$

$$G^j_{h,1} = (r^i)^2(\dot{y}_{8,n_o}T^j_n - \dot{y}_{4,n_o}W^j_n) + \rho^o r^i W_{n_o}R^j_n - r^i y_{8,n_o}H^j_n +$$

$$+ (n_o - 1)(n - l)\mu^o W_{n_o}T^j_n + 2\mu^o[H_{n_o} + (n_o l - n_o - l)T_{n_o})]W^j_n$$

$$+ (n_o + n + l)\mu^o[(l - n)W_{n_o}T^j_n - (n_o + l - n - 1)T_{n_o}W^j_n]$$

$$K^j_{h,2} = (r^i)^2\{-\dot{H}_{n_o}y^j_{2,n} - \frac{1}{4\pi G}\dot{R}_{n_o}(y^j_{6,n} - \frac{(n+1)}{r^i}R^j_n) -$$

$$-[\frac{1}{2}(n_o + n + l + 1)(n_o + n - l) - n_o n](\dot{T}_{n_o}y^j_{4,n} + \dot{W}_{n_o}y^j_{8,n})\}$$

$$G^j_{h,2} = (r^i)^2(\dot{W}_{n_o}y^j_{4,n} - \dot{T}_{n_o}y^j_{8,n})$$

$$K^j_{h,3} = \frac{1}{4\pi G}\{(n_o + 1)(n_o + 2) - \frac{(4n_o + 2)}{(1 + k_{n_o})}\}R_{n_o}R^j_n - (n + 1)\rho^o r^i H_{n_o}R^j_n$$

$$G^j_{h,3} = 0.$$

All the notations used here have been explained in the respective chapters.

**Appendix D. Relaxation of the Shear Modulus in the Earth's Mantle Based on Zschau's Relaxation Theory**

A new rheological model which describes the anelastic and viscous behaviour of the earth's mantle was recently developed by Zschau (personal communication, or see Zschau & Wang, 1986; Wang, 1986). He supposes a generalized Maxwell rheology with a cut Gaussian distribution of the logarithmic stress relaxation times. In determining the parameters of the distribution, a large set of data from various observations of inelasticity over a frequency range of more than seven decades including the seimological band, free oscillations, earth tides, the Chandler wobble, post glacial rebound as well as laboratory experiments has been used. The effect of composition, structure, phase transitions, pressure and temperature on the visco-elastic behaviour is accounted for by making use of the empirical relationship

(D-1) $$G^* = c\frac{\Phi}{\rho_o}$$

where $G^*$ is the Gibb's free energy of a relaxation process, $\Phi$ the seismic parameter, $\rho_o$ the ambient density at normal condition, and c a constant which is only dependent on the relaxation mechanism itself and assumed to be distributed.

The assumed cut Gaussian distribution of the logarithmic stress relaxation times is expressed by

(D-2) $$R(\ln\tau) = \begin{cases} R_o \exp\left[-\left(\frac{\ln(\tau/\tau_*)}{\sqrt{2}\,\sigma}\right)^2\right], & \tau < \tau_1; \\ 0, & \tau \geq \tau_1 \end{cases}$$

where $\ln\tau_*$ is the central, $\ln\tau_1$ the cut-off logarithmic relaxation time, and $\sigma$ the standard deviation for the corresponding full Gaussian distribution. From relationship (D-1), these quantities may be written as

(D-3) $$\begin{cases} \tau_* = \tau_o \exp\left(a_*\frac{\Phi}{\rho_o T}\right) \\ \tau_1 = \tau_o \exp\left(a_1\frac{\Phi}{\rho_o T}\right) \\ \sigma = b\frac{\Phi}{\rho_o T} \end{cases}$$

where $\tau_o$, $a_*$, $a_1$ and b are global constants and have been given by Zschau as

$$(\text{D-4})\quad \begin{cases} \tau_o = 3.0 \times 10^{-11} \text{ s}, \\ a_* = 200 \times 10^{-8} \text{ cgs}, \\ a_1 = 55 \times 10^{-8} \text{ cgs}, \\ b = 28.5 \times 10^{-8} \text{ cgs}, \qquad (1 \text{ cgs unit} = 1 \text{ g}\cdot\text{cm}^{-5}\cdot\text{s}^2\cdot\text{K}). \end{cases}$$

The constant $R_o$ is determined by the normalization condition

$$(\text{D-5})\qquad \int_{-\infty}^{\infty} R(\ln\tau)\,d(\ln\tau) = 1,$$

and given by

$$(\text{D-6})\qquad R_o = \sqrt{\frac{2}{\pi}}\,\frac{1}{\sigma\cdot \operatorname{erfc}(x_1)}$$

$$x_1 = \frac{\ln(\tau_*/\tau_1)}{\sqrt{2}\,\sigma}, \qquad \operatorname{erfc}(x_1) = \frac{2}{\sqrt{\pi}}\int_{x_1}^{\infty} \exp(-x^2)\,dx.$$

The relaxed shear modulus $\mu(s)$ in the (complex) frequency domain is determined by the integral

$$(\text{D-7})\qquad \mu(s) = \mu_o \int_{-\infty}^{\ln\tau_1} R(\ln\tau)\,\frac{s\tau}{1+s\tau}\,d(\ln\tau)$$

where $\mu_o$ is the unrelaxed or very high frequency shear modulus. This integral can only be solved numerically.

It can be shown that a direct method for solving this integral will be, in general, not stable because of very steep behaviour of the integrand. We therefore look for a series solution by making use of the expansion

$$(\text{D-8})\qquad \frac{s\tau}{1+s\tau} = \begin{cases} (s\tau)\sum\limits_{k=0}^{\infty}(-s\tau)^k, & |s\tau| < 1: \\ \sum\limits_{k=0}^{\infty}(-s\tau)^{-k}, & |s\tau| > 1. \end{cases}$$

Omitting the details, we give the series solution as follows

$$\text{(D-9)}\qquad \mu(s) = \begin{cases} \dfrac{\mu_o}{F(x_1)}\,(s\tau_1)\,\Sigma_F\!\left(s\tau_1,\ \dfrac{\sigma}{\sqrt{2}},\ \dfrac{\sigma}{\sqrt{2}} + x_1\right), & |s\tau_1| \le 1, \\[2ex] \dfrac{\mu_o}{F(x_1)}\left\{ \exp\!\left(x_1^2 - x_2^2\right)\left[(s\tau_2)\,\Sigma_F\!\left(s\tau_2,\ \dfrac{\sigma}{\sqrt{2}},\ \dfrac{\sigma}{\sqrt{2}} + x_2\right) + \right.\right. & \\ \left.\left. + \Sigma_F\!\left(\dfrac{1}{s\tau_2},\ \dfrac{\sigma}{\sqrt{2}},\ -x_2\right)\right] - \Sigma_F\!\left(\dfrac{1}{s\tau_1},\ \dfrac{\sigma}{\sqrt{2}},\ -x_1\right)\right\}, & |s\tau_1| > 1, \end{cases}$$

where

$$\text{(D-10)}\qquad x_2 = \frac{\ln(\tau_*/\tau_2)}{\sqrt{2}\,\sigma} \quad \text{and} \quad \tau_2 = \frac{1}{|s|}\,.$$

$\Sigma_F$ denotes series

$$\text{(D-11)}\qquad \Sigma_F\left(z,\ \alpha,\ \beta\right) = \sum_{k=0}^{\infty} (-z)^k F(k\alpha + \beta), \qquad |z| \le 1,\quad \mathrm{Re}(z) \ge 0,$$

and the function F is defined as

$$\text{(D-12)}\qquad F(x) = \exp\left(x^2\right)\cdot \mathrm{erfc}(x), \qquad F(-x) = 2\exp\left(x^2\right) - F(x).$$

This function can be very accurately evaluated by the approximation formulae

$$\text{(D-13)}\qquad F(x) = \exp\left(x^2\right)\sum_{k=1}^{5} c_k t^k + \varepsilon(x), \qquad \text{for } 0 \le x \ge 2,$$

$$t = \frac{1}{1 + px}, \qquad p = 0.3275911, \quad |\varepsilon(x)| \le 8.2 \times 10^{-6}.$$

$c_1 = 0.254829592$, $\qquad c_2 = -0.284496736$,

$c_3 = 1.421413741$, $\qquad c_4 = -1.453152027$,

$c_5 = 1.061405429$.

(Abramowitz & Stegun, 1964) and

(D-14) $$F(x) = \sum_{k=0}^{6} (-1)^k c_k x^{-2k-1} + \varepsilon(x), \qquad \text{for } x > 2,$$

$c_o = 0.5641896,$ $\qquad$ $c_1 = 0.2820442,$

$c_2 = 0.4198641,$ $\qquad$ $c_3 = 0.9688819,$

$c_4 = 2.3948264,$ $\qquad$ $c_5 = 4.4010273,$

$c_6 = 3.9054630,$ $\qquad$ $|\varepsilon(x)| \leq 1.0 \times 10^{-6}.$

For large x, function $F(x) = c_o/x + O(x^{-3})$. This property ensures the convergence of all series in (D-9) for any complex Laplace parameter s ($Re(s) > 0$) or Fourier frequency $s = i\omega$.

## References

Abramowitz, M., and Stegun, I. A., 1964: Handbook of Mathematical Functions, *Applied Mathematics Series* **55**, National Bereau of Standards, Washington, reprinted 1968 by Dover, New York.

Alterman, Z., Jarosch, H., and Pekeris, C.L., 1959: Oscillations of the Earth. *Proc. R. Soc. London A* **252**, 80-95.

Anderson, O.L., Schreiber, E., Liebermann, R.C., and Soga, N., 1968: Some elastic constant data on minerals relevant to geophysics. *Rev. Geophys. Space Phys.*, **6**, 491-524.

Anderson, D.L., and Minster, J.B, 1979: The frequency dependence of Q in the Earth and implications for mantle rheology and Chandler wobble. *Geophys. J. R. Astron. Soc.*, **58**, 431-440.

Backus, G.E., 1967: Converting vector and tensor equations to scalar equations in spherical coordinates. *Geophys. J. R. Astron. Soc.*, **13**, 71.

Biot, M.A., 1954: Theory of stress-strain relations in anisotropic viscosity relation phenomena. *J. Appl. Phys.* **25**, No. 11, 1385-1391.

Carter, W.E., Robertson, D.S., and MacKay, J.R., 1985: Geodetic Radio Interferometric Surveying: Applications and Results. *J. Geophys. Res.* **90**, 4577-4587.

Christodoulidis, D.C., Smith, D.E., Klosko, S.M., and Dunn, P.J., 1986: Solid Earth and Ocean Tide Parameters from LAGEOS. In *Proc. 10-th Int. Symp. Earth Tides*, edited by R. Vieira, Consejo Superior de Investigaciones Cientificas, Madrid, Spain, pp. 953-961.

Dahlen, F.A., 1972: Elastic dislocation theory of a self gravitating elastic configuration with an initial static stress field. *Geophys. J. R. Astron. Soc.*, **28**, 357-383.

Dehant, V., 1987: Integration of the gravitational motion equations for an elliptical uniformly rotating Earth with an inelastic mantle. *Phys. Earth Planet. Int.*, **49**, 242-258.

Dehant, V., and Ducarme, B., 1987: Comparison between the Theoretical and Observed Tidal Gravimetric Factors. *Phys. Earth Plant. Inter.* **49**, 192-212.

Dehant, V., and Zschau, J., 1989: The effect of mantle inelasticity on tidal gravity: a comparison between the spherical and elliptical Earth model; *Geophys. J.*, **97** (3), 549-555.

Dziewonski, A.M., Hager, B.H., and O'Connell, R.J., 1977: Large-scale heterogeneities in the lower mantle. *J. Geophys. Res.*, **82**, 239-255.

Dziewonski, A.M., 1984: Mapping the Lower Mantle: Determination of Lateral Heterogeneity in P Velocity up to Degree and Order 6. *J. Geophys. Res.* **89**, No. B7, 5929-5952.

Dziewonski, A.M., and Anderson, D.L., 1981: Preliminary Reference Earth Model. *Phys. Earth Planet. Inter.* **25**, 297-356.

Edmonds, A.R., 1960: Angular Momentum in Quantum Mechanics, Princeton University Press, Princeton, New Jersey.

Farrell, W.E., 1972: Deformation of the Earth by surface loads. *Rev. Geophys. Space Phys.* **10**, 761-797.

Gilbert, F., and Backus, G., 1968: Elastic-gravitational vibrations of a radially stratified sphere, in *Dynamics of Stratified Solids*, pp. 82, ed. Herrmann, G., American Society of Mechanical Engineers, New York.

Jeffreys, H., and Vicente, R.O., 1957a: The Theory of Nutation and the Variation of Latitude. *Mon. Notic. Roy. Astron. Soc.* **117**, 142-161.

Jeffreys, H., and Vicente, R.O., 1957b: The Theory of Nutation and the Variation of Latitude - The Roche Model-Core. *Mon. Notic. Roy. Astron. Soc.* **117**, 162-173.

Kanamori, H., and Anderson, D.L., 1977: Importence of physical dispersion in surface-wave and free-oscillation problems, *Rev. Geophys. Space Phys.* **15**, 105.

Lamor, J., 1909: The Relation of the Earth's Free Precessional Nutation to Its Resistance against Tidal Deformation. *Proc. R. Soc. London A.* **82**, 89-96.

Liu, H.-P., Anderson, D.L., and Kanamori, H., 1976: Velocity dispersion due to anelasticity: Implications for seismology and mantle composition, *Geophys. J. R. Astron. Soc.* **47**, 41.

Longman, J.M., 1962: A Green's Function for Determining the Deformation of the Earth under Surface Mass Loads I: Theory. *J. Geophys. Res.* **67**, 845-850.

Longman, J.M., 1963: A Green's Function for Determining the Deformation of the Earth under Surface Mass Loads, II: Computations and Numerical Results. *J. Geophys. Res.* **68**, 485.

Love, A.E.H., 1909: The Yielding of the Earth to Disturbing Forces. *Proc. R. Soc. London A* **82**, 73-88.

Love, A.E.H., 1911: Some Problems of Geodynamics. Dover, New York.

Marsh, J.G., Lerch, F.J., Putney, B.H., Christodoulidis, D.C., Smith, D.E., Felsentreger, T.L., Sanchez, B.V., Klosko, S.M., Pavlis, E.C., Martin, T.V., Robbins, J.W., Williamson, R.G., Colombo, O.L., Rowlands, D.D., Eddy, W.F., Chnadler, N.L., Rachlin,K.E., Patel, G.B., Bhati, S., and Chinn, D.S., 1988: A New Gravitational Model of the Earth from Satellite Tracking Data: GEM-T1. *J. Geophys. Res.* **93**, No. B6, 6169-6215.

Masters, G., Jordan, T.H., Silver, P.G., and Gilbert, F., 1982: Aspherical earth structure from fundamental spheroidal-mode data. *Nature*, **298**, 609-613.

Minster, J.B., 1980: Anelasticity and attenuation. In *Proc. of the Intern. School of Physcs 'Enrico Fermi'. Physics of the Earth Interior*, edited by Dziewonski, A.M., and Boschi, E., pp. 277.

Molodenskij, S.M., 1961: Theory of Nutation and the Diurnal Earth Tides. Coll. (Sb) "Earth Tides and Nutation of the Earth", Acad. Sci. USSR Press, Moscow.

Molodenskij, S.M., 1977: The Influence of Horizontal Inhomgeneities in the mantle on the Amplitude of Tidal Oscillations. *Bull. (Izv.) Acad. Sci., USSR, Earth Physics* **13**, No. 2, 77-80.

Molodenskij, S.M., 1980: Tides in a Asymmetric Earth. *Investigation of Earth Tides*, Nauka.

Molodenskij, S.M., and Kramer, M.V., 1980: The Influence of Large-Scale Horizontal Inhomogeneities in the Mantle on Earth Tides. *Bull. (Izv.) Acad. Sci., USSR, Earth Physics* **16**, No. 1, 1-11.

Müller, G., 1983: Rheological properties and velocity dispersion of a medium with power-law dependence of Q on frequency. *J. Geophys. Res.* **54**, 20-29.

Nakanishi, I., and Anderson, D.L., 1983: Measurements of mantle wave velocities and inversion for lateral heterogeneity and anisotropy, 1, Analysis of great circle phase velocities. *J. Geophys. Res.* **88**, 10267-10283.

Phinney, R.A., and Burridge, R., 1973: Representation of the elastic-gravitational excitation of a spherical Earth model by generalized spherical harmonics. *Geophys. J. R. Astron. Soc.* **34**, 451-487.

Richards, M.A., and Hager, B.H., 1984: Geoid anomalies in a dynamic earth. *J. Geophys. Res.* **89**, 5987-6002.

Ryan, J.W., Clark, T.A., Coates, R.J., Ma, C., Wildes, W.T., Gwinn, C.R., Herring, T.A., Shapiro, I.I., Corey, B.E., Counselman, C.C., Hinteregger, H.F., Rogers, A.E.E., Whitney, A.R., Knight, C.A., Vandenberg, N.R., Pigg, J.C., Schupler, B.R., and Rönnäng, B.O., 1986: Geodesy by Radio Interferometry: Determinations of Baseline Vector, Earth Rotation, and Solid Earth Tide Parameters with the Mark I Very Long Baseline Radio Interferometry System. *J. Geophys. Res.* **91** (B2), 1935-1946.

Sasao, T., Okubo, S., and Saito, M., 1980: A simple theory on dynamical effects of stratified fluid core upon rotational motion of the earth. *Proc. IAU Symp. No. 78, Nutation and the Earth's Rotation*, edited by Fedorov, E.P., Smith, M.L, and Bender, P.L., Kiev, May 1977.

Takeuchi, H., Saito, M., and Kobayashi, N., 1962: Statical deformations and free oscillations of a model earth. *J. Geophys. Res.* **67**, 1141.

Wahr, J.M., 1979: The Tidal Motion of a Rotating, Elliptical, Elastic and Oceanless Earth. Ph.D. Thesis, University of Colorado, pp. 216.

Wahr, J.M., 1981a: Body Tides on an Elliptical, Rotating, Elastic and Oceanless Earth. *Geophys. J. R. Astron. Soc.* **64**, 677-703.

Wahr, J.M., 1981b: The Forced Nutations of an Elliptical, Rotating, Elastic and Oceanless Earth. *Geophys. J. R. Astron. Soc.* **64**, 705-727.

Wahr, J.M., 1981c: A Normal Mode Expansion for the Forced Response of a Rotating Earth. *Geophys. J. R. Astron. Soc.* **64**, 651-675.

Wahr, J.M., 1982: Computing tides, nutations and tidally-induced variations in the earth's rotation rate for a rotating, elliptical earth. In *Lecture at the Third Intern. Summer School in the Mountains, on Geodesy and Global Geodynamics*, edited by Moritz, H., and Sünkel, H., Admont, pp. 327-379.

Wahr, J.M., Sasao, T., and Smith, M.L., 1981: Effect of the Fluid Core on Changes in the Length of Day due to Long Period Tides. *Geophys. J. R. Astron. Soc.* **64**, 635-650.

Wakker, K.F., Ambrosius, B.A.C., and Aardoom, L., 1985: Orbit Determination and European Station Positioning From Satellite Laser Ranging; *J. Geophys. Res.*, **90** (B11), 9275-9284.

Wang, R., 1986: Das viskoelastische Verhalten der Erde auf langfristige Gezeitenterme. *Diplomarbeit, Math.-Nat. Fak., Univ. Kiel.*

Wang, R., and Zschau, J., 1986: Tidal Effects of Temperature Induced Lateral Inhomogeneities in the Mantle. Joint Meeting of EGS and ESC, Kiel, 1986. Abstract in *Terra cognita* **6**, Vol. 3, 465.

Wiggins, R.A., 1968: Terrestrial variational tables for the periods and attenuation of the free oscillations, *Phys. Earth Planet. Int.*, **1**, 201.

Woodhouse, J.H., and Dziewonski, A.M., 1984: Mapping the Upper Mantle: Three-Dimensional Modeling of Earth Structure by Inversion of Seismic Waveforms. *J. Geophys. Res.* **89**, No. B7, 5953-5986.

Zschau, J., 1979: Auflastgezeiten. *Habilitationsschrift, Math.-Nat. Fak. Univ. Kiel.*

Zschau, J., 1983: Rheology of the Earth's mantle at tidal and Chandler wobble periods; in J.T. Kuo, *9th Int. Symp. Earth Tides*, New York, 1981, 605-630, Schweizerbart'sche Verlagsbuchhandlung, Stuttgart.

Zschau, J., 1986: Tidal friction in the solid Earth: Constraints from the Chandler wobble period; in A.J. Anderson and A. Cazenave, *Space Geodesy and Geodynamics*, 316-344. Academinc Press, London.

Zschau, J., and Wang, R., 1986: Imperfect Elasticity in the Earth's Mantle: Implications for the Earth Tides and Long Period Deformations. in *Proc. 10-th Int. Symp. Earth Tides* edited by R. Vieira, Conseja Superior de Investigaciones Cientificas, Madrid, Spain, pp 379-384.

## Acknowledgements

The successful completion of the research presented in this thesis would not have been possible without the help of a number of my colleagues and friends.

Firstly, I would like to express my sincere gratitude to my supervisor, Prof. Dr. J. Zschau, for suggestion and persistent guidance of this thesis. His constructive criticism and useful comments played an important role in this work. The heated discussions with him about many special problems were very valuable and are still very vivid in my impression.

My deepest thanks are due to Prof. Dr. R. Meißner for showing concern for this work, and due to many colleagues of the Institute of Geophysics, Kiel, especially the members of the "Geodynamics Group" for various supports. I am greatly indepted to Dr. H.-P. Plag for careful reading of the manuscript and for useful comments and corrections.

I also wish to express my profound thanks to Prof. H.-T. Hsu for support and critical discussions, and to my wife Cuixia Chen and many Chinese friends, whose various helps in some way or the other contributed to my studies in Germany in no small way.

I also acknowledge the financial support of the Christian-Albrechts-University, Kiel.

Rainer Uhrenbacher

## A New Method for Interpreting Tectonomagnetic Field Changes Using a Natural Geomagnetic Stress Sensor

A Contribution to the Joint German-Turkish Earthquake Prediction Research Project

Frankfurt/M., Bern, New York, Paris, 1989. 226 pp.
European University Studies: Series 17, Earth Sciences. Vol. 4
ISBN 3-631-41817-5 pb. DM 59.--/sFr. 50.--

The author introduces a new method to determine crustal stress changes from tectonomagnetic field variations. The principal part of the new approach is a structural model of a geological formation with stress dependent magnetic rock properties, which acts as a "natural geomagnetic stress sensor". The new technique was applied to data obtained from the investigation area of the German-Turkish Earthquake Prediction Research Project at the western end of the North-Anatolian Fault Zone. Here, the differential magnetic total intensity field is being measured biannually since Mai 1985. The author compares the stress changes determined from tectonomagnetic field variations with the results obtained from other experiments and demonstrates that the new method could be significant for monitoring crustal stress changes.

*Contents:* Natural geomagnetic stress sensor – magneto-elastic rock mechanics – tectonomagnetic field changes – interpretation of field observations

**Verlag Peter Lang Frankfurt a.M. · Bern · New York · Paris**
Auslieferung: Verlag Peter Lang AG, Jupiterstr. 15, CH-3000 Bern 15
Telefon (004131) 321122, Telex pela ch 912 651, Telefax (004131) 321131
- Preisänderungen vorbehalten -

Zeitfracht Medien GmbH
Ferdinand-Jühlke-Straße 7
99095 Erfurt, Deutschland
produktsicherheit@kolibri360.de

Druck:
CPI Druckdienstleistungen GmbH
im Auftrag der
Zeitfracht Medien GmbH
Ein Unternehmen der Zeitfracht - Gruppe
Ferdinand-Jühlke-Str. 7
99095 Erfurt